SpringerBriefs in Electrical and Computer Engineering

T0242339

For further volumes:
http://www.springer.com/series/10059

Leonardo Rey Vega · Hernan Rey

A Rapid Introduction to Adaptive Filtering

 Springer

Leonardo Rey Vega
School of Engineering
University of Buenos Aires
Paseo Colón 850
C1063ACV
Buenos Aires
Argentina

Hernan Rey
Department of Engineering
University of Leicester
University Road
Leicester
LE1 7RH
UK

ISSN 2191-8112   ISSN 2191-8120   (electronic)
ISBN 978-3-642-30298-5   ISBN 978-3-642-30299-2   (eBook)
DOI 10.1007/978-3-642-30299-2
Springer Heidelberg New York Dordrecht London

Library of Congress Control Number: 2012939513

Printed on acid-free paper

Springer is part of Springer Science+Business Media (www.springer.com)

Preface

This book is intended to be a quick reference on basic concepts of adaptive filtering for students, practising engineers, and researchers. In order to use it for a graduate level course on adaptive filtering, it should be complemented with some introductory background material and more importantly, with exercises and computer projects. Therefore, it is important that the reader has some familiarity with basic concepts from probability theory and stochastic processes (first and second order statistics, correlation, stationarity, ergodicity, etc.), algebra, matrix analysis, and linear systems (stability, linearity, etc.)

We wanted to do more than just derive algorithms and perform simulations showing their behavior. We prioritized to present the material under a solid theoretical background. However, after showing several different interpretations associated to each algorithm, we included some intuitive thoughts in order to improve the understanding of how they operate. We believe this is important since equations, as exact and important as they are, can sometimes obscure some simple and fundamental aspects of an algorithm. We also developed a convergence analysis using standard and non-standard approaches. Theoretical results are important since they allow us to make predictions about the performance of an algorithm, which is very interesting at the design stage when solving a particular problem. In addition, we covered a few applications so the reader can see how adaptive filters can be used in solving real-world problems. Several references are included so the interested reader can look for more in-depth discussions about several important topics in adaptive filtering.

We would like to thank Sara Tressens for giving us the opportunity to take our first steps into the world of scientific research while we were still young undergraduates. We are also grateful to Prof. Jacob Benesty for allowing us to develop our ideas and encourage us to write this book. Finally, we thank especially Dr. Veronica Mas for proofreading the manuscript and providing helpful comments.

Paris, France, March 2012   Leonardo Rey Vega
Leicester, UK, March 2012   Hernan Rey

Contents

Acronyms

APA	Affine Projection Algorithm
BER	Bit Error Rate
EMSE	Excess Mean Square Error
EWRLS	Exponentially Weighted RLS
LHS	left hand side
LMS	Least Mean Squares
LS	Least Squares
MAE	Mean Absolute Error
MMSE	Minimum Mean Square Error
MSD	Mean Square Deviation
MSE	Mean Square Error
NLMS	Normalized Least Mean Squares
NR	Newton-Raphson
pdf	probability density function
RHS	right hand side
RLS	Recursive Least Squares
SD	Steepest Descent
SDA	Sign-Data Algorithm
SEA	Sign-Error Algorithm
SNR	Signal to Noise Ratio
SSA	Sign-Sign Algorithm
WLS	Weighted Least Squares
i.i.d	independent and identically distributed
$\mathbb{R}^L$	Real L-dimensional vector space
$\mathbb{C}^L$	Complex L-dimensional vector space
$\langle \mathbf{a}, \mathbf{b} \rangle$	Inner product between vectors $\mathbf{a}$ and $\mathbf{b}$
$\mathrm{eig}_i[\mathbf{A}]$	ith eigenvalue of matrix $\mathbf{A}$
$\mathrm{eig}_{\max}[\mathbf{A}]$	Maximum eigenvalue of matrix $\mathbf{A}$
$\mathrm{eig}_{\min}[\mathbf{A}]$	Minimum eigenvalue of matrix $\mathbf{A}$
$\chi(\mathbf{A})$	Condition number of matrix $\mathbf{A}$

A^{-1}	Inverse of matrix $\mathbf{A}$
$A^{\dagger}$	Moore-Penrose pseudoinverse of matrix $\mathbf{A}$
$\mathcal{N}(\mathbf{A})$	Null space of matrix $\mathbf{A}$
$\mathcal{R}(\mathbf{A})$	Range or column space of matrix $\mathbf{A}$
Rank $(\mathbf{A})$	Rank of matrix $\mathbf{A}$
$E[X]$ or $E_X[X]$	Expectation of random variable X
$E[X|Y]$ or $E_{X|Y}[X|Y$	Conditional expectation of X given Y

Chapter 1
Introduction

Abstract The area of adaptive filtering is a very important one in the vast field of Signal Processing. Adaptive filters are ubiquitous in current technology. System identificaction, equalization for communication systems, active noise cancellation, speech processing, sonar, seismology, beamforming, etc, are a few examples from a large set of applications were adaptive filters are used to solve different kinds of problems. In this chapter we provide a short introduction to the adaptive filtering problem and to the different aspects that should be taken into account when choosing or designing an adaptive filter for a particular application.

1.1 Why Adaptive Filters?

The area of adaptive filtering has experienced a great growth in the last 50 years, and even today it is constantly expanding. This allowed for the possibility of applying adaptive filters in such a variety of applications as system identification, echo cancelation, noise cancelation, equalization, control, deconvolution, change detection, smart antennas, speech processing, data compression, radar, sonar, biomedicine, seismology, etc. [1, 2]. Certainly, this would not have been possible without the VLSI technology necessary to provide the computational requirements for these applications [3].

A filter becomes adaptive when its coefficients are updated (using a particular algorithm) each time a new sample of the signals is available. Conceptually, the algorithm chosen should be able to follow the evolution of the system under study. This provides the adaptive filter with the ability to operate in real time and improves its performance with no need to have any users involved [2].

Figure 1.1 shows a scheme of an adaptive filtering problem. Basically, the adaptive filter receives a vector input signal $\mathbf{x}(n)$ and a reference/desired signal $d(n)$.[1] At time

---

[1] Unless otherwise stated, we focus in this book on finite impulse response (FIR) filters, and random signals being time-discrete, real-valued, and zero-mean.

L. Rey Vega and H. Rey, *A Rapid Introduction to Adaptive Filtering*,
SpringerBriefs in Electrical and Computer Engineering,
DOI: 10.1007/978-3-642-30299-2_1, © The Author(s) 2013

Fig. 1.1 Scheme of an adaptive filtering problem

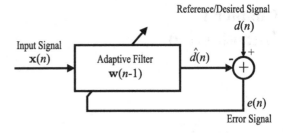

n, the filter coefficients $\mathbf{w}(n-1) = [w_0(n-1), w_1(n-1), \ldots, w_{L-1}(n-1)]^T$, with $(\cdot)^T$ denoting transposition, will be updated according to a specific algorithm.

As we will see in Chap. 2, Wiener filtering provides the linear optimum filtering solution (in mean-square-error sense). However, we will also see that the associated cost function and its optimal solution require knowledge of the input autocorrelation matrix and the cross-correlation vector. These statistics capture the information required about the system under study. Then, we may ask: Why should we seek adaptive algorithms? There are indeed several reasons, e.g.:

- *Savings in memory*. In practice, the statistics are not available, so they need to be estimated. This computation could require the accumulation of a lot of samples from the input and desired signals.
- *Introduction of delay*. The accumulation and post-processing introduce delay in the filter output. This issue is critical in certain applications like echo cancelation, adaptive delay estimation, low-delay predictive coding, noise cancelation, radar, and channel equalization applications in mobile telephony.
- *No a priori statistical information required*. The adaptive algorithm learns the statistics from only one realization of the input and reference processes. Hence, the designer has no need to know a priori the underlying signal statistics.[2]
- *Tracking ability*. If the statistics (of the input and/or reference signals) change with time, the optimal filter will also change, so it would require its recalculation. The learning mechanism of the adaptive algorithm can track these changes in the statistics. Again, different algorithms will do this at different rates and with different degrees of success.
- *Ease of implementation*. In general, they are easy to implement, making them suitable for real-world applications.

Three basic items are important in the design of an adaptive filter [3]:

1. *Definition of the objective function form*. Some examples for the objective function are Mean Square Error (MSE), Least Squares (LS), Weighted Least Squares (WLS), Instantaneous Squared Value (ISV) and Mean Absolute Error (MAE). The MSE, in a strict sense, is only of theoretical value, since it requires an

---

[2] Even if some statistical information is available, it can be used combined with the input and reference realization.

"infinite amount" of signal samples to be computed (given it has the expectation operator in its definition). In practice, this ideal objective function can be approximated by the LS, WLS, or ISV. They differ in the implementation complexity and in the convergence behavior characteristics. In general, the ISV is easier to implement but it presents noisy convergence properties, since it represents a greatly simplified objective function. The LS is convenient to be used in stationary environments, whereas the WLS is useful in applications where the environment is slowly varying or when the measurement noise is not white. MAE is particularly useful in the presence of outliers and leads to computational simple algorithms.

2. *Definition of the error signal.* The choice of the error signal is crucial for the algorithm definition, since it can affect several of its characteristics, including computational complexity, speed of convergence, and robustness.

3. *Definition of the minimization algorithm.* This item is the main subject of Optimization Theory, and it essentially affects the speed of convergence and computational complexity of the adaptive process. As we will see in Chap. 3, Taylor approximations of the cost function with respect to the update can be used. The first-order approximation leads to the gradient search algorithm, whereas if the second-order approximation is used, Newton's algorithm arises (if the Hessian matrix is not positive definite, then the function does not necessarily have a unique minimum). Quasi-Newton's algorithms are used when the Hessian is estimated, but they are susceptible to instability problems due to the recursive form used to generate the estimate of the inverse Hessian matrix. In these methods, the step size controls the stability, speed of convergence, and some characteristics of the residual error of the overall iterative process.

Several properties should be considered to determine whether or not a particular adaptive filtering algorithm is suitable for a particular application. Among them, we can emphasize speed of convergence, steady state error, tracking ability, robustness against noise, computational complexity, modularity, numerical stability and precision [1]. Describing these features for a given adaptive filter can be a very hard task from a mathematical point of view. Therefore, certain hypotheses like linearity, Gaussianity and stationarity have been used over the years to make the analysis more tractable [4]. We will show this in more detail in the forthcoming chapters.

1.2 Organization of the Work

In the next chapter we start studying the problem of optimum linear filtering with stationary signals (particularly, in the mean-square-error sense). This leads to the Wiener filter. Different values for the filter coefficients would lead to larger mean square errors, a relation that is captured by the error performance surface. We will study this surface as it will be very useful for the subsequent chapters. We also include and example where the Wiener filter is applied to the linear prediction problem.

In Chap. 3 we will introduce iterative search methods for minimizing cost functions. This will be particularly applied to the error performance surface. We will focus on the methods of Steepest Descent and Newton-Raphson, which belong to the family of deterministic gradient algorithms. These methods uses the second order statistics to find the optimal filter (i.e., the Wiener filter) but unlike the direct way introduced in Chap. 2, they find this solution iteratively. The iterative mechanism will lead to the question of stability and convergence behavior, which we will study theoretically and with simulations. Understanding their functioning and convergence properties is very important as they will be the basis for the development of stochastic gradient adaptive filters in Chap. 4

Chapter 4 presents the Stochastic Gradient algorithms. Among them, we find the Least Mean Square algorithm (LMS), which is without doubt the adaptive filter that has been used the most. Other algorithms studied in this chapter are the Normalized Least Mean Square (NLMS), the Sign Data algorithm (SDA), the Sign Error algorithm (SEA), and the Affine Projection algorithm (APA). We show several interpretations of each algorithm that provide more insight on their functioning and how they relate to each other. Since these algorithms are implemented using stochastic signals, the update directions become subject to random fluctuations called gradient noise. Therefore, a convergence analysis to study their performance (in statistical terms) will be performed. This will allow us to generate theoretical predictions in terms of the algorithms stability and steady state error. Simulation results will be provided to test this predictions and improve the understanding of the pros and cons of each algorithm. In addition, we discuss the applications of adaptive noise cancelation and adaptive equalization.

In Chap. 5 we will study the method of Least Squares (LS). In this case, a linear regression model is used for the data and the estimation of the system using input-output measured pairs (and no statistical information) is performed. As the LS problem can be thought as one of orthogonal projections in Hilbert spaces, interesting properties can be derived. Then, we will also present the Recursive Least Squares (RLS) algorithm, which is a recursive and more computational efficient implementation of the LS method. We will discuss some convergence properties and computational issues of the RLS. A comparison of the RLS with the previously studied adaptive filters will be performed. The application of adaptive beamforming is also developed using the RLS algorithm.

The final chapter of the book deals with several topics not covered in the previous chapters. These topics are more advanced or are the object of active research in the area of adaptive filtering. We include a succinct discussion of each topic and provide several relevant references for the interested reader.

References

1. S. Haykin, *Adaptive Filter Theory*, 4th edn. (Prentice-Hall, Upper Saddle River, 2002)
2. D.G. Manolakis, V.K. Ingle, S.M. Kogon, *Statistical and Adaptive Signal Processing* (Artech House, Norwood, 2005)
3. P.S.R. Diniz, *Adaptive Filtering: Algorithms And Practical Implementation*, 3rd edn. (Springer, Boston, 2008)
4. S. Haykin, Signal processing: where physics and mathematics meet. IEEE Signal Process. Mag. **18**, 6–7 (2001)

Chapter 2
Wiener Filtering

Abstract Before moving to the actual adaptive filtering problem, we need to solve the optimum linear filtering problem (particularly, in the mean-square-error sense). We start by explaining the analogy between linear estimation and linear optimum filtering. We develop the principle of orthogonality, derive the Wiener–Hopf equation (whose solution lead to the optimum Wiener filter) and study the error surface. Finally, we applied the Wiener filter to the problem of linear prediction (forward and backward).

2.1 Optimal Linear Mean Square Estimation

Lets assume we have a set of samples $\{x(n)\}$ and $\{d(n)\}$ coming from a jointly wide sense stationary (WSS) process with zero mean. Suppose now we want to find a linear estimate of $d(n)$ based on the L-most recent samples of $x(n)$, i.e.,

$$\hat{d}(n) = \mathbf{w}^T \mathbf{x}(n) = \sum_{l=0}^{L-1} w_l x(n-l), \quad \mathbf{w}, \mathbf{x}(n) \in \mathbb{R}^L \quad \text{and} \quad n = 0, 1, \dots \quad (2.1)$$

The introduction of a particular criterion to quantify how well $d(n)$ is estimated by $\hat{d}(n)$ would influence how the coefficients w_l will be computed. We propose to use the *Mean Squared Error* (MSE), which is defined by

$$J_{\text{MSE}}(\mathbf{w}) = E\left[|e(n)|^2\right] = E\left[|d(n) - \hat{d}(n)|^2\right], \quad (2.2)$$

where $E[\cdot]$ is the expectation operator and $e(n)$ is the estimation error. Then, the estimation problem can be seen as finding the vector $\mathbf{w}$ that minimizes the cost function $J_{\text{MSE}}(\mathbf{w})$. The solution to this problem is sometimes called the *stochastic*

L. Rey Vega and H. Rey, *A Rapid Introduction to Adaptive Filtering*,
SpringerBriefs in Electrical and Computer Engineering,
DOI: 10.1007/978-3-642-30299-2_2, © The Author(s) 2013

least squares solution, which is in contrast with the deterministic solution we will study in Chap. 5

If we choose the MSE cost function (2.2), the optimal solution to the linear estimation problem can be presented as:

$$\mathbf{w}_{\text{opt}} = \arg \min_{\mathbf{w} \in \mathbb{R}^L} J_{\text{MSE}}(\mathbf{w}). \tag{2.3}$$

Replacing (2.1) in (2.2), the latter can be expanded as

$$J_{\text{MSE}}(\mathbf{w}) = E\left[|d(n)|^2 - 2d(n)\mathbf{x}(n)^T \mathbf{w} + \mathbf{w}^T \mathbf{x}(n)\mathbf{x}^T(n)\mathbf{w}\right]. \tag{2.4}$$

As this is a quadratic form, the optimal solution will be at the point where the cost function has zero gradient, i.e.,

$$\nabla_{\mathbf{w}} J_{\text{MSE}}(\mathbf{w}) = \frac{\partial J_{\text{MSE}}}{\partial \mathbf{w}} = \mathbf{0}_{L \times 1}, \tag{2.5}$$

or in other words, the partial derivative of J_{MSE} with respect to each coefficient w_l should be zero.

2.2 The Principle of Orthogonality

Using (2.1) in (2.2), we can compute the gradient as

$$\frac{\partial J_{\text{MSE}}}{\partial \mathbf{w}} = 2E\left[e(n)\frac{\partial e(n)}{\partial \mathbf{w}}\right] = -2E\left[e(n)\mathbf{x}(n)\right]. \tag{2.6}$$

Then, at the minimum,[1] the condition that should hold is:

$$E\left[e_{\min}(n)\mathbf{x}(n)\right] = \mathbf{0}_{L \times 1}, \tag{2.7}$$

or equivalently

$$E\left[e_{\min}(n)x(n-l)\right] = 0, \quad l = 0, 1, \ldots, L-1. \tag{2.8}$$

This is called the *principle of orthogonality*, and it implies that the optimal condition is achieved if and only if the error $e(n)$ is decorrelated from the samples $x(n-l)$, $l = 0, 1, \ldots, L-1$. Actually, the error will also be decorrelated from the estimate $\hat{d}(n)$ since

---

[1] The Hessian matrix of J_{MSE} is positive definite (in general), so the gradient of J_{MSE} is nulled at its minimum.

Fig. 2.1 Illustration of the
principle of orthogonality
for $L = 2$. The optimal
error $e_{\min}(n)$ is orthogonal to
the input samples $x(n)$ and
$x(n-1)$, and to the optimal
estimate $\hat{d}_{\mathrm{opt}}(n)$. It should
be noticed that the notion of
orthogonality in this chapter
is equivalent to the notion of
decorrelation

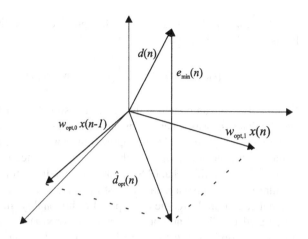

$$E\left[e_{\min}(n)\hat{d}_{\mathrm{opt}}(n)\right] = E\left[e_{\min}(n)\mathbf{w}_{\mathrm{opt}}^T\mathbf{x}(n)\right] = \mathbf{w}_{\mathrm{opt}}^T E\left[e_{\min}(n)\mathbf{x}(n)\right] = 0. \quad (2.9)$$

Fig. 2.1 illustrates the orthogonality principle for the case $L = 2$.

2.3 Linear Optimum Filtering

Consider a signal $x(n)$ as the input to a finite impulse response (FIR) filter of length
L, $\mathbf{w}_T = [w_{T,0}, w_{T,1}, \ldots, w_{T,L-1}]^T$. This filtering operation generates an output

$$y(n) = \mathbf{w}_T^T \mathbf{x}(n), \quad (2.10)$$

with $\mathbf{x}(n) = [x(n), x(n-1), \ldots, x(n-L+1)]^T$. As the output of the filter is
observed, it can be corrupted by an additive measurement noise $v(n)$, leading to a
linear regression model for the observed output

$$d(n) = \mathbf{w}_T^T \mathbf{x}(n) + v(n). \quad (2.11)$$

It should be noticed that this linear regression model can also be used even if the
input-output relation of the given data pairs $[\mathbf{x}(n), d(n)]$ is nonlinear, with $\mathbf{w}_T$ being
a linear approximation to the actual relation between them. In that case, in $v(n)$ there
would be a component associated to the additive noise perturbations, but also another
one representing, for example, modeling errors.

In the context of (2.11), we can look at $\mathbf{w}_T$ as the quantity to be estimated by a linear
filter $\mathbf{w} \in \mathbb{R}^L$, with (2.1) giving the output of this filter. This output can still be seen
as an estimate of the reference signal $d(n)$ or the system's output $y(n)$. Therefore,
the problem of optimal filtering is analogous to the one of linear estimation.

When J_{MSE} is the cost function to be optimized, the orthogonality principle (2.7) holds, which can be put as:

$$E\left[e_{\min}(n)\mathbf{x}(n)\right] = E\left\{\left[d(n) - \mathbf{w}_{\text{opt}}^T\mathbf{x}(n)\right]\mathbf{x}(n)\right\} = \mathbf{0}_{L \times 1}, \qquad (2.12)$$

From (2.12) we can conclude that given the signals $x(n)$ and $d(n)$, we can always assume that $d(n)$ was generated by the linear regression model (2.11). To do this, the system $\mathbf{w}_T$ would be equal to the optimal filter $\mathbf{w}_{\text{opt}}$, while $v(n)$ would be associated to the residual error $e_{\min}(n)$, which will be uncorrelated to the input $\mathbf{x}(n)$ [1].

It should be noticed that (2.8) is not just a condition for the cost function to reach its minimum, but also a mean for testing whether a linear filter is operating in the optimal condition. Here, the principle of orthogonality illustrated in Fig. 2.1 can be interpreted as follows: at time n the input vector $\mathbf{x}(n) = [x(n), x(n-1)]^T$ will pass through the optimal filter $\mathbf{w}_{\text{opt}} = [w_{\text{opt},0}, w_{\text{opt},1}]^T$ to generate the output $\hat{d}_{\text{opt}}(n)$. Given $d(n)$, $\hat{d}_{\text{opt}}(n)$ is the only element in the space spanned by $\mathbf{x}(n)$ that leads to an error $e(n)$ that is orthogonal to $x(n)$, $x(n-1)$, and $\hat{d}_{\text{opt}}(n)$.[2]

2.4 Wiener–Hopf Equation

Now we focus on the computation of the optimal solution. From (2.12), we have

$$E\left[\mathbf{x}(n)\mathbf{x}^T(n)\right]\mathbf{w}_{\text{opt}} = E\left[\mathbf{x}(n)d(n)\right]. \qquad (2.13)$$

We introduce the following definitions

$$\mathbf{R}_{\mathbf{x}} = E\left[\mathbf{x}(n)\mathbf{x}^T(n)\right] \quad \text{and} \quad \mathbf{r}_{\mathbf{x}d} = E\left[\mathbf{x}(n)d(n)\right] \qquad (2.14)$$

for the input autocorrelation matrix and the cross correlation vector, respectively. Note that as the joint process is WSS, the matrix $\mathbf{R}_{\mathbf{x}}$ is symmetric, positive definite[3] and Toeplitz [2]. Using these definitions, equation (2.13) can be put as

$$\mathbf{R}_{\mathbf{x}}\mathbf{w}_{\text{opt}} = \mathbf{r}_{\mathbf{x}d}. \qquad (2.15)$$

This is the compact matrix form of a set of L equations known as *Wiener–Hopf equations* and provides a way for computing the optimal filter (in MSE sense) based on

---

[2] In this chapter we use the idea of orthogonality as a synonym of decorrelation. This orthogonality we are referring to is given in a certain space of random variables with finite variance and an inner product between x and y defined by $E[xy]$.

[3] The autocorrelation matrix is certainly positive semidefinite. For it not to be positive definite, some linear dependencies between the random variable conforming $\mathbf{x}(n)$ would be required. However, this is very rare in practice.

some statistical properties of the input and reference processes. Under the assumption on the positive definiteness of $\mathbf{R_x}$ (so that it will be nonsingular), the solution to (2.15) is:

$$\mathbf{w}_{\text{opt}} = \mathbf{R_x}^{-1}\mathbf{r}_{\mathbf{x}d}, \tag{2.16}$$

which is known as the *Wiener filter*. An alternative way to find it is the following. Using the definitions (2.14) into (2.4) results in

$$J_{\text{MSE}}(\mathbf{w}) = E\left[|d(n)|^2\right] - 2\mathbf{r}_{\mathbf{x}d}^T\mathbf{w} + \mathbf{w}^T\mathbf{R_x}\mathbf{w}. \tag{2.17}$$

In addition, it can be easily shown that the following factorization holds:

$$\mathbf{w}^T\mathbf{R_x}\mathbf{w} - 2\mathbf{r}_{\mathbf{x}d}^T\mathbf{w} = (\mathbf{R_x}\mathbf{w} - \mathbf{r}_{\mathbf{x}d})^T\,\mathbf{R_x}^{-1}\,(\mathbf{R_x}\mathbf{w} - \mathbf{r}_{\mathbf{x}d}) - \mathbf{r}_{\mathbf{x}d}^T\mathbf{R_x}^{-1}\mathbf{r}_{\mathbf{x}d}. \tag{2.18}$$

Replacing (2.18) in (2.17) leads to

$$J_{\text{MSE}}(\mathbf{w}) = E\left[|d(n)|^2\right] - \mathbf{r}_{\mathbf{x}d}^T\mathbf{R_x}^{-1}\mathbf{r}_{\mathbf{x}d} + \left(\mathbf{w} - \mathbf{R_x}^{-1}\mathbf{r}_{\mathbf{x}d}\right)^T\mathbf{R_x}\left(\mathbf{w} - \mathbf{R_x}^{-1}\mathbf{r}_{\mathbf{x}d}\right). \tag{2.19}$$

Using the fact that $\mathbf{R_x}$ is positive definite (and therefore, so is its inverse), it turns out that the cost function reaches its minimum when the filter takes the form of (2.16), i.e., the Wiener filter. The minimum MSE value (MMSE) on the surface (2.19) is:

$$J_{\text{MMSE}} = J_{\text{MSE}}(\mathbf{w}_{\text{opt}}) = E\left[|d(n)|^2\right] - \mathbf{r}_{\mathbf{x}d}^T\mathbf{R_x}^{-1}\mathbf{r}_{\mathbf{x}d} = E\left[|d(n)|^2\right] - E\left[|\hat{d}_{\text{opt}}(n)|^2\right]. \tag{2.20}$$

We could have also arrived to this result by noticing that $e_{\min}(n) = d(n) - \hat{d}_{\text{opt}}(n)$ and using the orthogonality principle as in (2.9). Therefore, the MMSE is given by the difference between the variance of the reference signal $d(n)$ and the variance of its optimal estimate $\hat{d}_{\text{opt}}(n)$.

It should be noticed that if the signals $x(n)$ and $d(n)$ are orthogonal ($\mathbf{r}_{\mathbf{x}d} = 0$), the optimal filter will be the null vector and $J_{\text{MMSE}} = E\left[|d(n)|^2\right]$. This is reasonable since nothing can be done with the filter $\mathbf{w}$ if the input signal carries no information about the reference signal (as they are orthogonal). Actually, (2.17) shows that in this case, if any of the filter coefficients is nonzero, the MSE would be increased by the term $\mathbf{w}^T\mathbf{R_x}\mathbf{w}$, so it would not be optimal. On the other hand, if the reference signal is generated by passing the input signal through a system $\mathbf{w_T}$ as in (2.11), with the noise $v(n)$ being uncorrelated from the input $x(n)$, the optimal filter will be

$$\mathbf{w}_{\text{opt}} = \mathbf{R_x}^{-1}\mathbf{r}_{\mathbf{x}d} = \mathbf{R_x}^{-1}E\left\{\mathbf{x}(n)\left[\mathbf{x}^T(n)\mathbf{w_T} + v(n)\right]\right\} = \mathbf{w_T}. \tag{2.21}$$

This means that the Wiener solution will be able to identify the system $\mathbf{w_T}$ with a resulting error given by $v(n)$. Therefore, in this case $J_{\text{MMSE}} = E\left[|v(n)|^2\right] = \sigma_v^2$.

Finally, it should be noticed that the autocorrelation matrix admits the eigendecomposition:

$$\mathbf{R_x} = \mathbf{Q}\boldsymbol{\Lambda}\mathbf{Q}^T, \tag{2.22}$$

with $\boldsymbol{\Lambda}$ being a diagonal matrix determined by the eigenvalues $\lambda_0, \lambda_1, \ldots, \lambda_{L-1}$ of $\mathbf{R_x}$, and $\mathbf{Q}$ a (unitary) matrix that has the associated eigenvectors $\mathbf{q}_0, \mathbf{q}_1, \ldots, \mathbf{q}_{L-1}$ as its columns [2]. Lets define the *misalignment vector* (or weight error vector)

$$\tilde{\mathbf{w}} = \mathbf{w}_{\text{opt}} - \mathbf{w}, \tag{2.23}$$

and its transformed version

$$\mathbf{u} = \mathbf{Q}^T \tilde{\mathbf{w}}. \tag{2.24}$$

Using (2.20), (2.16), (2.23), (2.22), and (2.24) in (2.19), results in

$$J_{\text{MSE}}(\mathbf{w}) = J_{\text{MMSE}} + \mathbf{u}^T \boldsymbol{\Lambda} \mathbf{u}. \tag{2.25}$$

This is called the *canonical form* of the quadratic form $J_{\text{MSE}}(\mathbf{w})$ and it contains no cross-product terms. Since the eigenvalues are non-negative, it is clear that the surface describes an elliptic hyperparaboloid, with the eigenvectors being the principal axes of the hyperellipses of constant MSE value.

2.5 Example: Linear Prediction

In the filtering problem studied in this chapter, we use the L-most recent samples $x(n), x(n-1), \ldots, x(n-L+1)$ and estimate the value of the reference signal at time n. The idea behind a *forward linear prediction* is to use a certain set of samples $x(n-1), x(n-2), \ldots$ to estimate (with a linear combination) the value $x(n+k)$ for $k \geq 0$. On the other hand, in a *backward linear prediction* (also known as *smoothing*) the set of samples $x(n), x(n-1), \ldots, x(n-M+1)$ is used to linearly estimate the value $x(n-k)$ for $k \geq M$.

2.5.1 Forward Linear Prediction

Firstly, we explore the forward prediction case of estimating $x(n)$ based on the previous L samples. Since $\mathbf{x}(n-1) = [x(n-1), x(n-2), \ldots, x(n-L)]^T$, using a transversal filter $\mathbf{w}$ the forward linear prediction error can be put as

$$e_{f,L}(n) = x(n) - \sum_{j=1}^{L} w_j x(n-j) = x(n) - \mathbf{w}^T \mathbf{x}(n-1). \tag{2.26}$$

To find the optimum forward filter $\mathbf{w}_{f,L} = [w_{f,1}, w_{f,2}, \ldots, w_{f,L}]^T$ we minimize the MSE. The input correlation matrix would be

$$E\left[\mathbf{x}(n-1)\mathbf{x}^T(n-1)\right] = E\left[\mathbf{x}(n)\mathbf{x}^T(n)\right] = \mathbf{R_x} = \begin{bmatrix} r_x(0) & r_x(1) & \cdots & r_x(L-1) \\ r_x(1) & r_x(0) & \cdots & r_x(L-2) \\ \vdots & \vdots & \ddots & \vdots \\ r_x(L-1) & r_x(L-2) & \cdots & r_x(0) \end{bmatrix},$$

(2.27)

where $r_x(k)$ is the autocorrelation function for lag k of the WSS input process. As for the cross correlation vector, the desired signal would be $x(n)$, so

$$\mathbf{r}_f = E\left[\mathbf{x}(n-1)x(n)\right] = [r_x(1), r_x(2), \ldots, r_x(L)]^T. \tag{2.28}$$

As $\mathbf{w}_{f,L}$ will be the Wiener filter, it satisfies the modified Wiener–Hopf equation

$$\mathbf{R_x}\mathbf{w}_{f,L} = \mathbf{r}_f. \tag{2.29}$$

In addition, we can use (2.20) to write the forward prediction error power

$$P_{f,L} = r_x(0) - \mathbf{r}_f^T\mathbf{w}_{f,L}. \tag{2.30}$$

Actually, (2.29) and (2.30) can be put together into the *augmented Wiener–Hopf equation* as:

$$\begin{bmatrix} r_x(0) & \mathbf{r}_f^T \\ \mathbf{r}_f & \mathbf{R_x} \end{bmatrix} \mathbf{a}_L = \begin{bmatrix} P_{f,L} \\ \mathbf{0}_{L\times1} \end{bmatrix}, \tag{2.31}$$

where $\mathbf{a}_L = \begin{bmatrix} 1 & -\mathbf{w}_{f,L}^T \end{bmatrix}^T$. In fact the block matrix on the left hand side is the autocorrelation matrix of the $(L+1) \times 1$ input vector $[x(n), x(n-1), \ldots, x(n-L)]^T$. According to (2.26), when this vector passes through the filter $\mathbf{a}_L$ it produces the forward linear prediction error as its output. For this reason, $\mathbf{a}_L$ is known as the *forward prediction error filter*.

Now, in order to estimate $x(n)$ we might use only the $(L-i)$-most recent samples, leading to a prediction error

$$e_{f,L-i}(n) = x(n) - \sum_{j=1}^{L-i} w_j x(n-j). \tag{2.32}$$

But the orthogonality principle tells us that when using the optimum forward filter

$$E\left[e_{f,L}(n)\mathbf{x}(n-1)\right] = \mathbf{0}_{L\times1}. \tag{2.33}$$

Then, we can see that for $1 \le i \le L$,

$$E\left[e_{f,L}(n)e_{f,L-i}(n-i)\right] = E\left\{e_{f,L}(n)\mathbf{a}_{L-i}^T\left[x(n-i),\ldots,x(n-L)\right]^T\right\} = 0.$$
$$(2.34)$$

Therefore, we see that as $L \to \infty$, $E\left[e_f(n)e_f(n-i)\right] = 0$, which means that the sequence of forward errors $e_f(n)$ is asymptotically white. This means that a sufficiently long forward prediction error filter is capable of *whitening* a stationary discrete-time stochastic process applied to its input.

2.5.2 Backward Linear Prediction

In this case we start by trying to estimate $x(n-L)$ based on the next L samples, so the backward linear prediction error can be put as

$$e_{b,L}(n) = x(n-L) - \sum_{j=1}^{L} w_j x(n-j+1) = x(n-L) - \mathbf{w}^T\mathbf{x}(n). \qquad (2.35)$$

To find the optimum backward filter $\mathbf{w}_{b,L} = [w_{b,1}, w_{b,2}, \ldots, w_{b,L}]^T$ we minimize the MSE. Following a similar procedure as before to solve the Wiener filter, the augmented Wiener–Hopf equation has the form

$$\begin{bmatrix} \mathbf{R_x} & \mathbf{r}_b \\ \mathbf{r}_b^T & r_x(0) \end{bmatrix} \mathbf{b}_L = \begin{bmatrix} \mathbf{0}_{L\times 1} \\ P_{b,L} \end{bmatrix}, \qquad (2.36)$$

where $\mathbf{r}_b = E[\mathbf{x}(n)x(n-L)] = [r_x(L), r_x(L-1), \ldots, r_x(1)]^T$, $P_{b,L} = r_x(0) - \mathbf{r}_b^T\mathbf{w}_{b,L}$, and $\mathbf{b}_L = \left[-\mathbf{w}_{b,L}^T \ 1\right]^T$ is the *backward prediction error filter*.

Consider now a stack of backward prediction error filters from order 0 to L. If we compute the errors $e_{b,i}(n)$ for $0 \le i \le L$, it leads to

$$\mathbf{e}_b(n) = \begin{bmatrix} e_{b,0}(n) \\ e_{b,1}(n) \\ e_{b,2}(n) \\ \vdots \\ e_{b,L-1}(n) \end{bmatrix} = \begin{bmatrix} 1 & \mathbf{0}_{1\times(L-1)} \\ \mathbf{b}_1^T & \mathbf{0}_{1\times(L-2)} \\ \mathbf{b}_2^T & \mathbf{0}_{1\times(L-3)} \\ \vdots & \vdots \\ -\mathbf{w}_{b,L-1}^T & 1 \end{bmatrix} \mathbf{x}(n) = \mathbf{T}_b\mathbf{x}(n). \qquad (2.37)$$

The $L \times L$ matrix $\mathbf{T}_b$, which is defined in terms of the backward prediction error filter coefficients, is lower triangular with 1's along its main diagonal. The transformation (2.37) is known as *Gram–Schmidt orthogonalization* [3], which defines a one-to-one correspondence between $\mathbf{e}_b(n)$ and $\mathbf{x}(n)$.[4]

In this case, the principle of orthogonality states that

---

[4] The Gram–Schmidt process is also used for the orthogonalization of a set of linearly independent vectors in a linear space with a defined inner product.

$$E\left[e_{b,i}(n)x(n-k)\right] = 0 \qquad 0 \le k \le i-1. \tag{2.38}$$

Then, it is easy to show that, at each time n, the sequence of backward prediction errors of increasing order $\{e_{b,i}(n)\}$ will be decorrelated. This means that the autocorrelation matrix of the backward prediction errors is diagonal. More precisely,

$$E\left[e_b(n)e_b^T(n)\right] = \text{diag}\{P_{b,i}\} \qquad 0 \le i \le L-1. \tag{2.39}$$

Another way to get to this result comes from using (2.37) to write

$$E\left[e_b(n)e_b^T(n)\right] = T_b R_x T_b^T. \tag{2.40}$$

By definition, this is a symmetric matrix. From (2.36), it is easy to show that $R_x T_b^T$ is a lower triangular matrix with $P_{b,i}$ being the elements on its main diagonal. However, since T_b is also a lower triangular matrix, the product of both matrices must retain the same structure. But it has to be also symmetric, and hence, it must be diagonal.

Moreover, since the determinant of T_b is 1, it is a nonsingular matrix. Therefore, from (2.39) and (2.40) we can put

$$R_x^{-1} = T_b^T \text{diag}\{P_{b,i}\}^{-1} T_b = \left(\text{diag}\{P_{b,i}\}^{-1/2} T_b\right)^T \text{diag}\{P_{b,i}\}^{-1/2} T_b. \tag{2.41}$$

This is called the *Cholesky decomposition* of the inverse of the autocorrelation matrix. Notice that the inverse of the autocorrelation matrix is factorized into the product of an upper and lower triangular matrices that are related to each other through a transposition operation. These matrices are completely determined by the coefficients of the backward prediction error filter and the backward prediction error powers.

2.6 Final Remarks on Linear Prediction

It should be noticed that a sufficiently long (high order) forward prediction error filter transforms a (possibly) correlated signal into a white sequence of forward errors (the sequence progresses with time index n). On the other hand, the Gram–Schmidt orthogonalization transforms the input vector $x(n)$ into an equivalent vector $e_b(n)$, where its components (associated to the order of the backward prediction error filter) are uncorrelated.

By comparing the results shown for forward and backward predictions, it can be seen that: i) the forward and backward prediction error powers are the same. ii) the coefficients of the optimum backward filter can be obtained by reversing the ones of the optimum forward filter. Based on these relations, the *Levinson–Durbin algorithm*

[2] provides a mean of recurrently solving the linear prediction problem of order L with a complexity $O(L^2)$ instead of $O(L^3)$.[5]

2.7 Further Comments

The MSE defined in (2.2) uses the linear estimator $\hat{d}(n)$ defined in (2.1). If we relax the linear constraint on the estimator and look for a function of the input, i.e., $\hat{d}(n) = g(\mathbf{x}(n))$, the optimal estimator in mean square sense is given by the conditional expectation $E[d(n)|\mathbf{x}(n)]$ [4]. Its calculation requires knowledge of the joint distribution between $d(n)$ and $\mathbf{x}(n)$, and in general, it is a nonlinear function of $\mathbf{x}(n)$ (unless certain symmetry conditions on the joint distribution are fulfilled, as it is the case for Gaussian distributions). Moreover, once calculated it might be very hard to implement it. For all these reasons, linear estimators are usually preferred (which as we have seen, depend only on second order statistics).

On a historical note, Norbert Wiener solved a continuous-time prediction problem under causality constraints by means of an elegant technique now known as the Wiener–Hopf factorization technique. This is a much more complicated problem than the one presented in 2.3. Later, Norman Levinson formulated the Wiener filter in discrete time.

It should be noticed that the orthogonality principle used to derive the Wiener filter does not apply to FIR filters only; it can be applied to IIR (infinite impulse response) filtering, and even noncausal filtering. For the general case, the output of the noncausal filter can be put as

$$\hat{d}(n) = \sum_{i=-\infty}^{\infty} w_i x(n-i). \qquad (2.42)$$

Then, minimizing the mean square error leads to the Wiener–Hopf equations

$$\sum_{i=-\infty}^{\infty} w_{\text{opt},i} r_x(k-i) = r_{xd}(k), \quad -\infty < k < \infty \qquad (2.43)$$

which can be solved using Z-transform methods [5]. In addition, a general expression for the minimum mean square error is

---

[5] We use the Landau notation in order to quantify the computational cost of a numerical operation [3]. Assume that the numerical cost or memory requirement of an algorithm are given by a positive function $f(n)$, where n is the problem dimensionality. The notation $f(n) = O(g(n))$, where $g(n)$ is a given positive function (usually simpler than $f(n)$), means that there exists constants $M, n_0 > 0$ such that:

$$f(n) \le M g(n), \quad \forall n \ge n_0$$

.

$$J_{\text{MMSE}} = r_d(0) - \sum_{i=-\infty}^{\infty} w_{\text{opt},i} r_{xd}(i) \tag{2.44}$$

From this general case, we can derive the FIR filter studied before (index i in the summation and lag k in (2.43) go from 0 to $L - 1$) and the causal IIR filter (index i in the summation and lag k in (2.43) go from 0 to ∞).

Finally we would like to comment on the stationarity of the processes. We assume the input and reference processes were WSS. If this were not the case, the statistics would be time-dependent. However, we could still find the Wiener filter at each time n as the one that makes the estimation error orthogonal to the input, i.e., the principle of orthogonality still holds. A less costly alternative would be to recalculate the filter for every block of N signal samples. However, nearly two decades after Wiener's work, Rudolf Kalman developed the *Kalman filter*, which is the optimum mean square linear filter for nonstationary processes (evolving under a certain state space model) and stationary ones (converging in steady state to the Wiener's solution).

References

1. A.H. Sayed, *Adaptive Filters* (John Wiley & Sons, Hoboken, 2008)
2. S. Haykin, *Adaptive Filter Theory*, 4th edn. (Prentice-Hall, Upper Saddle River, 2002)
3. G.H. Golub, C.F. van Loan, *Matrix Computations* (The John Hopkins University Press, Baltimore, 1996)
4. B.D.O. Anderson, J.B. Moore, *Optimal Filtering* (Prentice-Hall, Englewood Cliffs, 1979)
5. T. Kailath, A.H. Sayed, B. Hassibi, *Linear estimation* (Prentice-Hall, Upper Saddle River, 2000)

Chapter 3
Iterative Optimization

Abstract In this chapter we introduce iterative search methods for minimizing cost functions, and in particular, the J_{MSE} function. We focus on the methods of Steepest Descent and Newton-Raphson, which belong to the family of deterministic gradient algorithms. Although these methods still require knowledge of the second order statistics as does the Wiener filter, they find this solution iteratively. We also study the convergence of both algorithms and include simulation results to provide more insights on their performance. Understanding their functioning and convergence properties is very important as they will be the basis for the development of stochastic gradient adaptive filters in the next chapter.

3.1 Introduction

In this chapter we present a set of algorithms that iteratively search for the minimum of the cost function. They do this based (at least) on the gradient of the cost function, so they are often called *deterministic gradient algorithms*. In order for the cost function to depend only on the filter $\mathbf{w}$, the statistics $\mathbf{R_x}$ and $\mathbf{r}_{\mathbf{x}d}$ must be given. In this way, these algorithms solve the Wiener-Hopf equation iteratively, most of them without requiring the (computationally challenging) inversion of the matrix $\mathbf{R_x}$. However, all the information from the environment is captured in the second-order statistics and these algorithms do not have a learning mechanism for adapting to changes in the environment. In the next chapter we will see how adaptive filters solve this issue. Particularly, the so called *stochastic gradient algorithms* will apply a deterministic gradient algorithm but with the true statistics being replaced by estimates. Therefore, understanding the development and convergence properties of the iterative methods we present in this chapter (relying on the known second order statistics) is very important for the study of adaptive algorithms (which rely on measured or estimated data).

L. Rey Vega and H. Rey, *A Rapid Introduction to Adaptive Filtering*,
SpringerBriefs in Electrical and Computer Engineering,
DOI: 10.1007/978-3-642-30299-2_3, © The Author(s) 2013

3.2 Steepest Descent Algorithm

Given a certain (differentiable) function $J(\mathbf{w})$, the optimal solution at its minimum will satisfy:

$$J(\mathbf{w}_{\text{opt}}) \le J(\mathbf{w}), \quad \forall \, \mathbf{w} \in \mathbb{R}^L. \tag{3.1}$$

The idea behind an iterative algorithm that searches for the optimal solution is to create a sequence $\mathbf{w}(n)$, starting from an initial condition $\mathbf{w}(-1)$, so that the associated values of the function lead to a decreasing sequence, i.e., $J(\mathbf{w}(n)) < J(\mathbf{w}(n-1))$.

Consider the following recursion:

$$\mathbf{w}(n) = \mathbf{w}(n-1) - \mu \mathbf{A} \nabla_{\mathbf{w}} J(\mathbf{w}(n-1)), \qquad \mathbf{w}(-1), \tag{3.2}$$

where $\mathbf{A}$ is a positive definite matrix and μ is a positive constant known as adaptation step size, that controls the magnitude of the update. If we assume a small value of μ, the update $\Delta \mathbf{w}(n) = \mathbf{w}(n) - \mathbf{w}(n-1)$ will be small. Then, we can do a first order Taylor series expansion around $\mathbf{w}(n-1)$,

$$J(\mathbf{w}(n)) = J(\mathbf{w}(n-1) + \Delta \mathbf{w}(n)) \approx J(\mathbf{w}(n-1)) + \nabla_{\mathbf{w}}^T J(\mathbf{w}(n-1)) \Delta \mathbf{w}(n), \tag{3.3}$$

which using (3.2) leads to

$$J(\mathbf{w}(n)) \approx J(\mathbf{w}(n-1)) - \mu \nabla_{\mathbf{w}}^T J(\mathbf{w}(n-1)) \mathbf{A} \nabla_{\mathbf{w}} J(\mathbf{w}(n-1)). \tag{3.4}$$

If the gradient is equal to the null vector, so is the update in (3.2), so the algorithm has converged to a stationary point of $J(\mathbf{w})$. In any other case, as $\mathbf{A}$ is positive definite, (3.4) indicates that $J(\mathbf{w}(n)) < J(\mathbf{w}(n-1))$. This shows that the recursion (3.2) will eventually converge to a minimum of the cost function (given that μ is "small" enough). It should be noticed that for an arbitrary function $J(\mathbf{w})$, the algorithm can converge to a local minimum rather than the global one (depending on the values chosen for the step size and the initial condition relative to local and global minima).

Now, we choose $\mathbf{A} = \mathbf{I}_L$, so the update is performed in the opposite direction of the gradient. With this choice, the recursion (3.2) is called *Steepest Descent* (SD) algorithm. From (3.3) and using Cauchy-Schwartz inequality, we see that

$$\frac{|J(\mathbf{w}(n)) - J(\mathbf{w}(n-1))|}{\|\Delta \mathbf{w}(n)\|} \le \|\nabla_{\mathbf{w}} J(\mathbf{w}(n-1))\|. \tag{3.5}$$

This means that the change rate of $J(\mathbf{w})$ from time $n-1$ to time n is upper bounded by $\|\nabla_{\mathbf{w}} J(\mathbf{w}(n-1))\|$, so the gradient gives the direction of maximum change rate, i.e., the direction of steepest descent. We should notice that before (3.3) we assumed a small update, so the direction of steepest descent is just a local property. As we will see later, this can affect the convergence of the algorithm, so that when approaching the minimum, the number of smaller updates needed to get an accurate solution might be increased.

Focusing now on the error surface J_{MSE}, remember from the previous chapter that we mentioned it has the shape of an elliptic hyperparaboloid. Therefore, it is intuitively reasonable that by successively moving in the opposite direction of the gradient of the surface, its minimum will be eventually reached since it has a unique global minimum. From (2.6), the gradient of J_{MSE} is:

$$\nabla_{\mathbf{w}} J_{MSE}(\mathbf{w}(n-1)) = -2E\left[\mathbf{x}(n)e(n)\right] = -2E\left[\mathbf{x}(n)d(n) - \mathbf{x}(n)\mathbf{x}^T(n)\mathbf{w}(n-1)\right]$$
$$= -2\left[\mathbf{r}_{\mathbf{x}d} - \mathbf{R}_{\mathbf{x}}\mathbf{w}(n-1)\right], \tag{3.6}$$

where in the last line we used the definitions of the second order statistics introduced in (2.14). Then, the SD algorithm can be put as

$$\mathbf{w}(n) = \mathbf{w}(n-1) + \mu\left[\mathbf{r}_{\mathbf{x}d} - \mathbf{R}_{\mathbf{x}}\mathbf{w}(n-1)\right], \qquad \mathbf{w}(-1), \tag{3.7}$$

where the factor 2 has been included into μ. Here, we see that if the algorithm converges, the second term of the right hand side will be zero, so the filter will lead to the Wiener solution given by (2.16). However, since the SD recursion can be understood in terms of a feedback model [1] (where $\mathbf{w}(n)$ and $\mathbf{w}(n-1)$ are linked through a feedback loop), the stability performance of the algorithm should be analyzed.

3.2.1 Convergence of the Steepest Descent Algorithm

Replacing (2.16) into (3.7), the SD recursion turns into

$$\mathbf{w}(n) = \mathbf{w}(n-1) + \mu\mathbf{R}_{\mathbf{x}}\left[\mathbf{w}_{opt} - \mathbf{w}(n-1)\right]. \tag{3.8}$$

Using the misalignment vector $\tilde{\mathbf{w}}(n)$ and its transformed version $\mathbf{u}(n)$ (defined in (2.23) and (2.24) respectively), the recursion can be written as

$$\mathbf{u}(n) = (\mathbf{I}_L - \mu\mathbf{\Lambda})\mathbf{u}(n-1). \tag{3.9}$$

This shows that under the coordinate system defined by the eigenvectors of $\mathbf{R}_{\mathbf{x}}$ and centered at $\mathbf{w}_{opt}$ there is no cross coupling, so each component of $\mathbf{u}(n)$ will converge independently. Actually, the dynamics for the i-th natural mode (with $i = 0, 1, \ldots, L-1$) are ruled by

$$u_i(n) = (1 - \mu\lambda_i)u_i(n-1), \tag{3.10}$$

that can be explicitly stated as

$$u_i(n) = (1 - \mu\lambda_i)^{n+1} u_i(-1). \tag{3.11}$$

Then, each eigenvalue λ_i determines the mode of convergence $(1 - \mu\lambda_i)$, which goes along the direction defined by its associated eigenvector. In order for the algorithm to converge as $n \to \infty$, the misalignment vector (and its transformed version) must vanish. Since (3.11) would be associated to an exponential behavior, the necessary and sufficient condition for the stability of the SD algorithm would be

$$|1 - \mu\lambda_i| < 1 \quad i = 0, 1, \ldots, L - 1. \tag{3.12}$$

This shows that the stability of the algorithm depends only on μ (a design parameter) and $\mathbf{R_x}$ (or more precisely, its eigenvalues). To satisfy the stability condition, the step size should be chosen according to

$$0 < \mu < \frac{2}{\lambda_{\max}}. \tag{3.13}$$

Recalling the canonical form of $J_{\mathrm{MSE}}(\mathbf{w})$ introduced in (2.25), we can use (3.11) to write

$$J_{\mathrm{MSE}}(n) = J_{\mathrm{MMSE}} + \xi(n) = J_{\mathrm{MMSE}} + \sum_{i=0}^{L-1} \lambda_i (1 - \mu\lambda_i)^{2(n+1)} u_i^2(-1). \tag{3.14}$$

The second term $\xi(n)$ is known as the *excess mean square error* (EMSE) and measures how far the algorithm is from the minimum. Equation (3.14) shows the evolution through the error surface as a function of the iteration number, and is known as the *learning curve* or MSE curve. It is the result of the sum of L exponentials associated to the natural modes of the algorithm. Since $\lambda_i(1 - \mu\lambda_i)^{2(n+1)} > 0 \ \forall i$, when (3.13) is satisfied the convergence is also monotonic (and EMSE goes to zero in steady state). Clearly, the choice of μ will not only affect the stability of the algorithm but also its convergence performance (when stable). Actually, from the L modes of convergence $(1 - \mu\lambda_i)$ there will be one with the largest magnitude, that will give the slowest rate of convergence to the associated component of the transformed vector $\mathbf{u}(n)$. Therefore, this will be the mode that determines the overall convergence speed of the SD algorithm. It is then possible to look for a value of μ that guarantees the maximum overall rate of convergence by minimizing the magnitude of the slowest mode, i.e.,

$$\mu_{\mathrm{opt}} = \arg\min_{\mu} \max_{\substack{i = 1, \ldots, L \\ |1 - \mu\lambda_i| < 1}} |1 - \mu\lambda_i|. \tag{3.15}$$

In looking for the μ_{opt} we only need to study the modes associated to $\lambda_{\max}$ and $\lambda_{\min}$ as a function of μ, since all the others will lie in between. For $\mu < \mu_{\mathrm{opt}}$ the mode associated to $\lambda_{\max}$ has smaller magnitude than the one associated to $\lambda_{\min}$. As the reverse holds for $\mu > \mu_{\mathrm{opt}}$, the optimal step size must satisfy the condition

$$1 - \mu_{opt}\lambda_{min} = -(1 - \mu_{opt}\lambda_{max}), \tag{3.16}$$

which leads to the solution

$$\mu_{opt} = \frac{2}{\lambda_{max} + \lambda_{min}}. \tag{3.17}$$

Although the two slowest modes will have the same speed, the mode $1 - \mu_{opt}\lambda_{min}$ will be positive, leading to an overdamped convergence where the trajectory to the minimum follows a continuous path. On the other hand, the negative mode $1 - \mu_{opt}\lambda_{max}$ leads to an underdamped convergence where the trajectory exhibits oscillations. If we use the ratio between the largest and smallest eigenvalues, which is known as the *condition number* (or *eigenvalue spread*) of $\mathbf{R_x}$ [2] and is denoted $\chi(\mathbf{R_x})$, the magnitude of the slowest modes can be put as

$$\alpha = \frac{\chi(\mathbf{R_x}) - 1}{\chi(\mathbf{R_x}) + 1}. \tag{3.18}$$

When $\lambda_{max} = \lambda_{min}$, $\alpha = 0$ so it takes one iteration to converge. As the condition number increases, so does α, becoming close to 1 for large condition numbers, which corresponds to a slow convergence mode. Therefore, $\chi(\mathbf{R_x})$ plays a critical role in limiting the convergence speed of the SD algorithm.

In practice, it is usual to choose μ in such a way that $\mu\lambda_i \ll 1$. Although this leads to a slower overdamped convergence, it might mitigate the effects that appear when the error surface is not fully known and needs to be measured or estimated from the available data (as it will be the case with stochastic gradient algorithms that will be studied in Chap. 4). Under this condition, the fastest mode is associated to the largest eigenvalue, while the slowest mode is associated to the smallest eigenvalue. Actually, since the step size bound is related to λ_{max}, the slowest mode associated to λ_{min} can be put as $1 - a/\chi(\mathbf{R_x})$, with $0 < a < 1$. Then, the larger the condition number, the slower the convergence of the algorithm. In any case, as long as the condition (3.13) is fulfilled, the SD algorithm applied to the error surface $J_{MSE}(\mathbf{w})$ will reach the global minimum (where the MSE takes the value J_{MMSE}) *regardless of the chosen initial condition*.

Finally, the step size can be turned into a time-varying sequence to improve the algorithm performance, e.g., choosing $\mu(n)$ to minimize $J_{MSE}(\mathbf{w}(n))$. However, the stability analysis needs to be revised. One possible sufficient condition for convergence would require the sequence $\{\mu_i\}$ to be square summable but not summable [3], e.g., $\mu_i = 1/(i + 1)$.

3.3 Newton-Raphson Method

Consider again the update (3.2). We showed that for any positive definite matrix $\mathbf{A}$, this recursion will converge to a minimum of the cost function. If we assume that the Hessian matrix of the cost function is positive definite, we can choose its inverse as the matrix $\mathbf{A}$, i.e.,

$$\mathbf{A} = \left[\nabla_{\mathbf{w}}^2 J_{\text{MSE}}(\mathbf{w}(n-1)) \right]^{-1}. \tag{3.19}$$

Then, the new recursion — starting with an initial guess $\mathbf{w}(-1)$ — is

$$\mathbf{w}(n) = \mathbf{w}(n-1) - \mu \left[\nabla_{\mathbf{w}}^2 J_{\text{MSE}}(\mathbf{w}(n-1)) \right]^{-1} \nabla_{\mathbf{w}} J_{\text{MSE}}(\mathbf{w}(n-1)). \tag{3.20}$$

This is known as the *Newton-Raphson* (NR) method since it is related to the method for finding the zeros of a function. To understand this relation, consider the univariate case where we want to find the zeros of a function $f(w)$, i.e., the solution to $f(w) = 0$. Starting with an initial guess w_0, define the next estimate w_1 as the point where the tangent at $f(w_0)$ intersects the w-axis. Then, as the derivative $f'(w_0)$ is the slope of the tangent at w_0, it holds

$$w_1 = w_0 - \frac{f(w_0)}{f'(w_0)}. \tag{3.21}$$

This will be applied recursively until we reach a zero of the function (so the update becomes null). Now, if we consider the function $f(w) = J'_{\text{MSE}}(w)$, when we find its zero we will find the minimum of $J_{\text{MSE}}(w)$. Extending the previous idea to the multidimensional case will lead to the NR recursion (3.20), where the step size μ is included as before to control the magnitude of the update.

In applying the NR method to $J_{\text{MSE}}(\mathbf{w})$, notice that from (3.6) the Hessian is given by $\mathbf{R_x}$, so (3.20) takes the form

$$\mathbf{w}(n) = \mathbf{w}(n-1) + \mu \mathbf{R_x}^{-1} \left[\mathbf{r}_{xd} - \mathbf{R_x} \mathbf{w}(n-1) \right] = \mathbf{w}(n-1) + \mu \left[\mathbf{w}_{\text{opt}} - \mathbf{w}(n-1) \right]. \tag{3.22}$$

The SD method performs the update in the direction of the gradient, so it is normal to the elliptic contour curves of the error surface. However, (3.22) shows that the NR method moves directly in the direction pointing towards the minimum. So the gradient is rotated as it is multiplied by $\mathbf{R_x}^{-1}$. To the gradient to be pointing directly towards the minimum, the point where the gradient is being evaluated must lie on the direction of one of the eigenvectors of $\mathbf{R_x}$, or the contour curves of the error surface need to be circular (so $\mathbf{R_x} \propto \mathbf{I}_L$).

Moreover, if we choose $\mu = 1$ the minimum is reached in one iteration of the NR algorithm regardless of the the initial guess. As true as this is, when the NR method is used in practice, errors in computation/estimation of the gradient and/or Hessian are very likely to arise. The use of $\mu < 1$ will give better control over the convergence of the algorithm. In any case, as it will be useful in the study of adaptive filters in the next chapter, it is important to see how this algorithm behave with a stable μ different from 1.

Subtracting $\mathbf{w}_{\text{opt}}$ from both sides of (3.22), we can write

$$\tilde{\mathbf{w}}(n) = (1 - \mu)\tilde{\mathbf{w}}(n - 1) = (1 - \mu)^{n+1}\tilde{\mathbf{w}}(-1). \tag{3.23}$$

In terms of its learning curve, using (3.23) into the canonical form (2.25) leads to

$$J_{\text{MSE}}(n) = J_{\text{MMSE}} + \tilde{\mathbf{w}}^T(n)\mathbf{R}_{\mathbf{x}}\tilde{\mathbf{w}}(n)$$
$$= J_{\text{MMSE}} + (1 - \mu)^{2(n+1)}[J_{\text{MSE}}(-1) - J_{\text{MMSE}}]. \tag{3.24}$$

So the NR algorithm is stable when $0 < \mu < 2$ (independently of $\mathbf{R}_{\mathbf{x}}$) and has only one mode of convergence (exponential and monotonic), depending entirely on μ. We have previously seen how slow modes arise in the SD algorithm because of the eigenvalue spread in $\mathbf{R}_{\mathbf{x}}$. The fact that $\mathbf{R}_{\mathbf{x}}$ is not affecting the convergence mode of the NR algorithm makes us believe that it will converge faster than the SD method (given that μ and other factors are left the same in both algorithms). This is true in general, and not only for quadratic cost functions. Sufficiently close to the minimum, all cost functions are approximately quadratic functions, so the NR algorithm can take a step straight to the minimum. Meanwhile, the magnitude of the gradient is very small, slowing down the SD performance. If we think that the Hessian is modifying the step size of the SD method, the NR algorithm takes more equidistant steps, while the SD algorithm takes huge steps where the gradient is large. As a consequence, the NR method might not perform so well initially if the initial guess is too far away from the minimum. Later, we will make further comments on how this can be overcome.

The price for using the NR algorithm is that the Hessian needs to be esti-mated/computed and also inverted. This is a computational intensive task and can become numerically unstable if special care is not taken. One thing that can be done is to modify the matrix $\mathbf{A}$ in (3.19) and add a *regularization* constant $\beta > 0$ to the Hessian matrix. This does not affect the positive definiteness of $\mathbf{A}$, so the algorithm would still move through the surface in a decreasing way. But the benefit is that this constant ensures that $\mathbf{A}$ is nonsingular and improves the conditioning of the Hessian matrix (a bad conditioned matrix, i.e., a matrix with large eigenvalue spread, will be close to singular, so computing its inverse is numerically challenging).

3.4 Example

In this section we perform numerical simulations to give more insight on the per-formance of the iterative methods introduced in this chapter. We use $L = 2$ so the minimum will be $\mathbf{w}_{\text{opt}} = [w_{\text{opt},0}, w_{\text{opt},1}]^T$. On each scenario, we construct the cor-relation matrix $\mathbf{R}_{\mathbf{x}}$ using (2.22), based on the eigenvalues and eigenvectors matrices

$$\Lambda = \begin{bmatrix} 1 & 0 \\ 0 & \frac{1}{\chi(\mathbf{R}_{\mathbf{x}})} \end{bmatrix} \qquad \mathbf{Q} = \frac{1}{\sqrt{2}}\begin{bmatrix} 1 & -1 \\ 1 & 1 \end{bmatrix}, \tag{3.25}$$

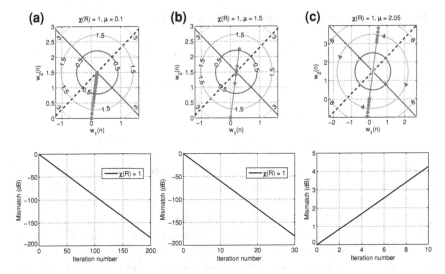

Fig. 3.1 Behavior of the steepest descent algorithm with $\chi(\mathbf{R_x}) = 1$. In the top plots showing the contour curves of the error surface and the trajectory of the algorithm, the principal axes (i.e., the eigenvectors of $\mathbf{R_x}$) are shown centered at $\mathbf{w}_{opt}$, with the dashed line representing the eigenvector associated to λ_{max}. The inial guess is $\mathbf{w}(-1) = \mathbf{0}$. **a** $\mu = 0.1$. **b** $\mu = 1.5$. **c** $\mu = 2.05$

where the conditioning number will be a parameter that will be varied. This means that $\lambda_{max} = 1$, so the step size μ needs to be chosen in the interval $(0, 2)$ to ensure stability. To simplify the description, we define each convergence mode for the SD algorithm as

$$\text{mode}_{max} = 1 - \mu\lambda_{max} = 1 - \mu, \quad \text{mode}_{min} = 1 - \mu\lambda_{min} = 1 - \mu/\chi(\mathbf{R_x}) \quad (3.26)$$

We will show the evolution of the estimate $\mathbf{w}(n)$ in relation to the contour plots of the error surface and we study the performance dynamics of the algorithm using the *mismatch*, which is defined as

$$10\log_{10}\frac{\|\mathbf{w}(n) - \mathbf{w}_{opt}\|^2}{\|\mathbf{w}_{opt}\|^2}. \quad (3.27)$$

We start with $\chi(\mathbf{R_x}) = 1$, which in this case means that $\mathbf{R_x} = \mathbf{I}_L$. In Fig. 3.1 we study the SD algorithm and use different step sizes to represent three different regimes: $0 < \mu < 1/\lambda_{max}$, $1/\lambda_{max} < \mu < 2/\lambda_{max}$ and $\mu > 2/\lambda_{max}$. As $\mathbf{R_x} = \mathbf{I}_L$, the contour curves are circles, so the direction normal to these curves (i.e., the direction of the gradient) points towards the minimum and the algorithm evolves following a straight line trajectory (i.e., at equal speed along both principal axes). On the left panel we see the case $\mu = 0.1$, so $\text{mode}_{max} = \text{mode}_{min} = 0.9$. Since the modes are positive, the convergence is overdamped so there is no change of sign in the transformed coordinate system. Even from the first iterations the algorithm takes small steps towards the minimum, which become even smaller as the iteration number progresses (since the magnitude of the gradient decreases). In the middle panel $\mu = 1.5$, so $\text{mode}_{max} = \text{mode}_{min} = -0.5$. These negative values lead to underdamped oscillations, so at each iteration it switches between two opposite quadrants in the transformed coordinate system (but it still does it along a straight line). Since these modes have a much smaller magnitude than in the previous scenario, the convergence speed is increased as it shows from comparing the mismatch between scenarios a) and b). In the right panel, $\mu = 2.05$, so $\text{mode}_{max} = \text{mode}_{min} = -1.05$. In this case it also alternates between quadrants following a straight line, but as it is unstable in both modes it moves further away from the optimum at each iteration. This can be corroborated with the monotonically increasing mismatch.

In Fig. 3.2 we change into the scenario $\chi(\mathbf{R_x}) = 2$. When $\mu = 0.1$, $\text{mode}_{max} = 0.9$ and $\text{mode}_{min} = 0.95$. The trajectory is no longer a straight line and since mode_{min} is closer to 1, the trajectory converges faster along the direction associated to λ_{max}. This means that it approaches faster to the direction of the eigenvector associated to λ_{min} (solid line). In addition, as the slowest mode is slower than the one in Fig. 3.1 (with the same step size), the mismatch show a slower convergence as well. In b), $\mu = 1.5$, so $\text{mode}_{max} = -0.5$ and $\text{mode}_{min} = 0.25$. On the one hand, we see that the fastest mode is associated to λ_{min}. On the other hand, the convergence will be overdamped in the direction associated to λ_{min} and underdamped in the other one. Both facts can be confirmed in the simulation. The algorithm approaches the minimum in a zigzag manner but it goes faster towards the dashed line (the direction associated to λ_{max}). Interestingly, the slowest mode has the same magnitude as in the analogous scenario from Fig. 3.1, so the mismatch in both cases is essentially the same. In the last case, $\mu = 2.5$, so $\text{mode}_{max} = -1.5$ and $\text{mode}_{min} = -0.25$. Although both modes are underdamped, one is stable and one is not. We see that the algorithm converges (quite quickly) in the direction associated to λ_{min} but then it ends up moving away from the minimum along the direction associated to λ_{max}, which is the unstable one. Overall, the algorithm is unstable and the mismatch will be divergent. With $\mu > 4$ the algorithm would be divergent in both directions.

The effect of increasing the eigenvalue spread to $\chi(\mathbf{R_x}) = 10$ is analyzed in Fig. 3.3. For $\mu = 0.1$, the resulting modes are $\text{mode}_{max} = 0.9$ and $\text{mode}_{min} = 0.99$. The speed difference between modes has been enlarged so the algorithm moves almost in an L-shape way, first along the direction of the fast mode (associated to λ_{max}) and finally along the slow mode direction. The overall convergence is clearly even slower than with the previous smaller condition numbers as shown in the mismatch curves. When $\mu = 1.5$, $\text{mode}_{max} = -0.5$ and $\text{mode}_{min} = 0.85$. The faster mode is underdamped and associated to λ_{max} while the slow mode is overdamped, so the algorithm moves quickly zigzagging along the direction of the slowest mode until it ends up moving slowly along it in an "almost" straight path to the minimum. Overall, the convergence is again slower than with the previous smaller condition numbers. Finally, $\mu = 2.05$ leads to $\text{mode}_{max} = -1.05$ and $\text{mode}_{min} = 0.795$. The mode associated to λ_{min} converges in an overdamped way, but since the other mode is unstable, the algorithm ends up moving away from the minimum along the direction

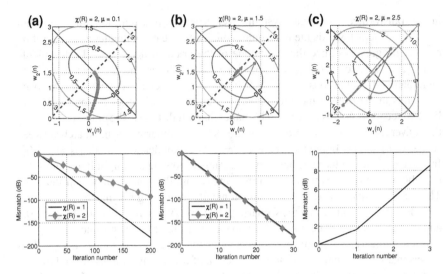

Fig. 3.2 Same as in Fig. 3.1 but with $\chi(\mathbf{R_x}) = 2$ and $\mu = 2.5$ in c). In the stable scenarios, the mismatch curves are being compared with the ones from previous $\chi(\mathbf{R_x})$ and using the same μ

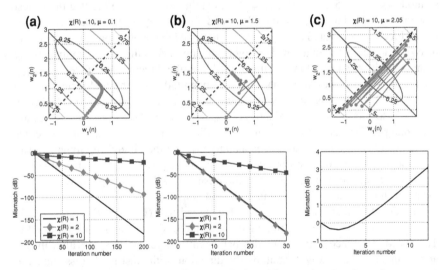

Fig. 3.3 Same as in Fig. 3.1 but with $\chi(\mathbf{R_x}) = 10$. In the stable scenarios, the mismatch curves are being compared with the ones from previous $\chi(\mathbf{R_x})$ and using the same μ

associated to λ_{max}. The fact that the magnitude of mode$_{min}$ is further away from 1 in comparison with the one of mode$_{max}$ makes the mismatch to decrease slightly in the first few iterations before the divergent mode becomes more prominent and causes the mismatch do increase monotonically.

In Fig. 3.4 we study two more scenarios with $\chi(\mathbf{R_x}) = 10$. Firstly, for $\mu = 0.8$ (mode$_{max} = 0.2$ and mode$_{min} = 0.92$), we see a more accentuated version of the

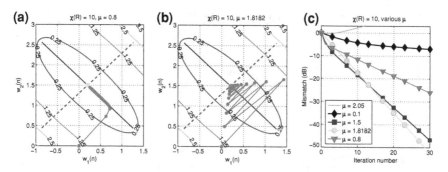

Fig. 3.4 Same as in Fig. 3.1 but with $\chi(\mathbf{R_x}) = 10$. **a** $\mu = 0.8$. **b** $\mu = \mu_{\text{opt}}$ given by (3.17). **c** Comparison of the mismatch for different μs

results shown in Fig. 3.3 a) as the mode associated to λ_{max} is much faster. Next, we test the optimal step size developed in (3.17), which in this scenario leads to $\mu_{\text{opt}} = 1.8182$. In this case, $\text{mode}_{\text{max}} = -0.8182$ and $\text{mode}_{\text{min}} = 0.8182$. As expected, both modes move at the same speed, but they converge in underdamped and overdamped ways, respectively. As it can be seen from the comparison of the mismatch curves, the one associated to μ_{opt} shows the smallest error after 30 iterations. In fact, if at each later iteration we sort the conditions in terms of decreasing mismatch, the ordering remains unchanged with respect to the one at iteration 30. However, in the first few iterations the case with $\mu = 0.8$ is the fastest, and to the iteration number 12, the mismatch for $\mu = 1.5$ is smaller than the one for μ_{opt}. It should be noticed that this is not inconsistent with the optimality of (3.17). We have to remember that the overall convergence of the algorithm is the result of adding the contributions of each mode. At the beginning of the iteration process, fast modes tend to give larger contributions than slower modes, which is reversed as we approach the minimum. Then, the result of Fig. 3.4 c) in the first few iterations is not surprising when we compare the fastest modes on each condition. The optimal step size μ_{opt} guarantees that in the later stages of convergence, as the slowest mode becomes dominant, the convergence will be the fastest relative to any other choice of the step size.

As we have seen in these examples, when μ is small enough so that all the modes are positive, it is true that the fastest and slowest modes will be associated to λ_{max} and λ_{min} respectively.

We finish this example using the NR method under the same scenario of Fig. 3.3. We know from (3.23) and (3.24) that the NR algorithm exhibits a single mode of convergence equal to $1 - \mu$, which is independent of $\chi(\mathbf{R_x})$. Since the scenarios used in Fig. 3.1 had $\chi(\mathbf{R_x}) = 1$ and $\lambda = 1$, we should expect to replicate those results when using the NR method. This is exactly what we see in Fig. 3.5. Moreover, the mismatch curves show that the results are unchanged when running the NR algorithm using different eigenvalues spread. See the next section for further comments on the NR and SD methods.

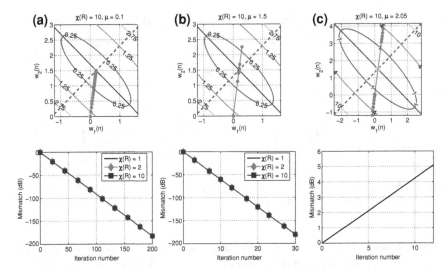

Fig. 3.5 Same as in Fig. 3.3 but for the Newton-Raphson algorithm

3.5 Further Comments

In [4], there is an interesting interpretation of the NR method that clarifies further why its convergence is independent of the eigenvalue spread. The result is that the NR algorithm works as an SD algorithm using an input signal generated by applying the Karhunen-Loéve transform (which decorrelates the input signal) and a power normalization procedure, which is known as a *whitening process*.

The SD method presents a very slow convergence rate in the vicinity of the optimal solution, which is overcome by the NR method. But the latter does not take much advantage of the high gradients at points far away from the minimum, as the SD method does. To improve this tradeoff, the Levenberg-Marquardt (LM) algorithm [5], comes as a combination of the SD and NR algorithms, trying to share the merits of both methods. It basically includes a time-dependent regularization constant $\beta(n)$ to the NR algorithm, and instead of multiplying it by the identity matrix $\mathbf{I}_L$ it uses the diagonal of the Hessian matrix. This last change allows each component of the gradient to be scaled independently to provide larger movement along the directions where the gradient is smaller. At the beginning of the iteration process, a large $\beta(n)$ is used, making the algorithm to move closer to the direction given by the gradient, i.e., operating as an SD algorithm. As we get closer to the minimum, $\beta(n)$ can be decreased, bringing the algorithm closer to the NR method. In addition the use of a regularization provides numerical stability to the NR method as mentioned earlier.

The attempts to reduce the computational cost of the NR method led to the family *quasi-Newton* methods [6]. The idea behind them is to approximate the Hessian matrix or its inverse in terms of the gradient. Some examples include the Broyden-Fletcher-Goldfarb-Shanno method (which as the LM method performs initially as an SD algorithm and later as an NR algorithm) and the Davidon-Fletcher-Powell method.

References

1. S. Haykin, *Adaptive Filter Theory*, 4th edn. (Prentice-Hall, Upper Saddle River, 2002)
2. G.H. Golub, C.F. van Loan, *Matrix Computations* (The John Hopkins University Press, Baltimore, 1996)
3. A.H. Sayed, *Adaptive Filters* (John Wiley & Sons, Hoboken, 2008)
4. B. Farhang-Boroujeny, *Adaptive Filters: Theory and Applications* (John Wiley & Sons, New York, 1998)
5. D. Marquardt, An Algorithm for Least-Squares Estimation of Nonlinear Parameters. SIAM Journal on Applied Mathematics, 11, 431–441 (1963).
6. C.-Y. Chi, C.-C. Feng, C.-H. Chen, C.-Y. Chen, *Blind Equalization and System Identification: Batch Processing Algorithms, Performance and Applications* (Springer, Berlin, 2006)

Chapter 4
Stochastic Gradient Adaptive Algorithms

Abstract One way to construct adaptive algorithms leads to the so called *Stochastic Gradient algorithms* which will be the subject of this chapter. The most important algorithm in this family, the Least Mean Square algorithm (LMS), is obtained from the SD algorithm, employing suitable estimators of the correlation matrix and cross correlation vector. Other important algorithms as the Normalized Least Mean Square (NLMS) or the Affine Projection (APA) algorithms are obtained from straightforward generalizations of the LMS algorithm. One of the most useful properties of adaptive algorithms is the ability of tracking variations in the signals statistics. As they are implemented using stochastic signals, the update directions in these adaptive algorithms become subject to random fluctuations called gradient noise. This will lead to the question regarding the performance (in statistical terms) of these systems. In this chapter we will try to give a succinct introduction to this kind of adaptive filter and to its more relevant characteristics.

4.1 Least Mean Square Algorithm

As mentioned in Sect. 3.1, approximating the statistics used in the gradient algorithms discussed in Chap. 3 leads to the family of adaptive filters known as *Stochastic Gradient algorithms*. As the statistics are computed based on the actual signals available, the resulting adaptive filters will have the potential to track their changes.

In 1960, Widrow and Hoff introduced the *Least Mean Squares* (LMS) algorithm. This was a milestone in the history of adaptive filtering. Almost 50 years later, it is fair to say that the LMS has been the most widely used and studied adaptive filter. The reasons for this are probably its simplicity (both in terms of memory requirements and computational complexity), stable and robust performance against different signal conditions, unbiased convergence in the mean to the Wiener solution, and stable behavior when implemented with finite-precision arithmetic. It has however some

L. Rey Vega and H. Rey, *A Rapid Introduction to Adaptive Filtering*,
SpringerBriefs in Electrical and Computer Engineering,
DOI: 10.1007/978-3-642-30299-2_4, © The Author(s) 2013

drawbacks, as it will be shown later, mostly related with a poor performance (slow convergence rate) with colored input signals.

4.1.1 Stochastic Gradient Approximation

Consider the Steepest Descent recursion in (3.7), rewritten here for ease of reference:

$$\mathbf{w}(n) = \mathbf{w}(n-1) + \mu \left[\mathbf{r}_{xd} - \mathbf{R}_x \mathbf{w}(n-1) \right], \ \mathbf{w}(-1). \tag{4.1}$$

The idea of the stochastic gradient approximation is to replace the correlation matrix and cross correlation vector by suitable estimates. The simplest approximation rely on the instantaneous values of the input and reference signals, i.e.,

$$\hat{\mathbf{R}}_x = \mathbf{x}(n)\mathbf{x}^T(n) \text{ and } \hat{\mathbf{r}}_{xd} = d(n)\mathbf{x}(n). \tag{4.2}$$

These estimates arise from dropping the expectation operator in the definitions of the statistics. Replacing the actual statistics in (4.1) by their estimates (4.2) leads to

$$\mathbf{w}(n) = \mathbf{w}(n-1) + \mu \mathbf{x}(n) e(n), \ \mathbf{w}(-1). \tag{4.3}$$

This is the recursion of the LMS or Widrow-Hoff algorithm. A common choice in practice is $\mathbf{w}(-1) = \mathbf{0}$.

As in the SD, the step size μ influences the dynamics of the LMS. It will be shown later that also as in the SD, large values of μ cause instability, whereas small ones give a low convergence rate. However, important differences should be stated. The filter $\mathbf{w}(n)$ in (4.3) is a random variable while the one in (4.1) is not. The MSE used by the SD is a deterministic function on the filter $\mathbf{w}$. The SD moves through that surface in the opposite direction of its gradient and eventually converges to its minimum. In the LMS, that gradient is approximated by

$$\hat{\nabla}_{\mathbf{w}} J \left(\mathbf{w}(n-1) \right) = \mathbf{x}(n) \left[\mathbf{w}^T(n-1)\mathbf{x}(n) - d(n) \right]. \tag{4.4}$$

On the other hand, we can think of (4.4) as the actual gradient of the instantaneous squared value (ISV) cost function, i.e.,

$$\hat{J} \left(\mathbf{w}(n-1) \right) = |e(n)|^2 = |d(n) - \mathbf{w}^T(n-1)\mathbf{x}(n)|^2, \tag{4.5}$$

where the factor 2 from the gradient calculation would be incorporated to the step size μ. This function arises from dropping the expectation in the definition of the MSE, and therefore it is now a random variable. At each time step, when new data is available, the shape of this cost function changes and the LMS moves in the opposite direction to its gradient.

This randomness makes the LMS update "noisy". Whereas the SD reaches the minimum of the error surface, the LMS will execute a random motion around the optimal point given by the Wiener filter. Although it might show convergence on average (which it does as it will be shown later), this motion will lead to a nonzero steady state EMSE, in contrast with the SD algorithm.

When μ is decreased, besides the reduction in convergence rate, the LMS decreases its EMSE. To understand the rationale behind this, consider the following idea. Even though the LMS is using instantaneous values as estimates for the true statistics, it is actually performing some averaging process on them during the adaptation given its recursive nature. When a small μ is used, the adaptation process progresses slowly, and the algorithm has a long "memory". The large amount of data allows the algorithm to learn the statistics better, leading to a performance (in terms of final MSE) closer to the one obtained by SD. To give more insight into this last idea, let $\mathbf{A}_{i,j}$ be defined as:

$$\mathbf{A}(i, j) = \begin{cases} \left[\mathbf{I}_L - \mu\mathbf{x}(i)\mathbf{x}^T(i)\right]\left[\mathbf{I}_L - \mu\mathbf{x}(i-1)\mathbf{x}^T(i-1)\right]\dots\left[\mathbf{I}_L - \mu\mathbf{x}(j)\mathbf{x}^T(j)\right] & i \geq j \\ \mathbf{I}_L & i < j \end{cases}$$
(4.6)

Now, after $N + 1$ iterations, the LMS estimate can be written as:

$$\mathbf{w}(N) = \mathbf{A}(N, 0)\mathbf{w}(-1) + \sum_{m=0}^{N} \mathbf{A}(N, m+1)\mu\mathbf{x}(m)d(m).$$
(4.7)

When $\mu \to 0$, the variance of the random variables $\mu\mathbf{x}(n)\mathbf{x}^T(n)$ and $\mu\mathbf{x}(n)d(n)$ will be negligible, and so the error in approximating their values by their expected values will be close to zero. Then, in this case,

$$\mathbf{w}(N) \approx (\mathbf{I}_L - \mu\mathbf{R_x})^{N+1}\mathbf{w}(-1) + \sum_{m=0}^{N} (\mathbf{I}_L - \mu\mathbf{R_x})^{N-m}\mu\mathbf{r_{xd}}.$$
(4.8)

Finally, when $N \to \infty$, the first term of (4.8) vanishes, and then,[1]

$$\lim_{N \to \infty} \mathbf{w}(N) \approx (\mu\mathbf{R_x})^{-1}\mu\mathbf{r_{xd}} = \mathbf{R_x}^{-1}\mathbf{r_{xd}},$$
(4.9)

which is the Wiener filter, and so the EMSE goes to zero.

---

[1] The convergence analysis will be properly done and justified in Sect. 4.5.3. Here we just want to give an intuitive result concerning the limiting behavior of the LMS.

4.1.2 Computational Complexity of LMS

Usually, when treating the computational complexity of a given numerical procedure, it is customary to consider only the computational cost of sums and products. These are by large the major contributors to the computational load of an algorithm and are almost independent of the digital platform where the algorithm is implemented. Therefore, they provide a first approximation of the algorithm complexity. A more precise account for the complexity of an algorithm should include static and dynamic memory usage and management, the use of floating or fixed point arithmetic, etc. As these could be platform dependent we do not consider them in the following analysis. The computational load per iteration of the LMS algorithm can be summarized as follows:

- Complexity of the filtering process: The filtering process basically consists in calculating the inner product $\mathbf{w}^T(n-1)\mathbf{x}(n)$. It is easy to see that this requires L multiplications and $L-1$ additions [1]. In order to compute $e(n)$ we need an extra addition, resulting in a total of L additions.
- Complexity of the update calculation: This include the computational load of obtaining $\mathbf{w}(n)$ from $\mathbf{w}(n-1)$, $\mathbf{x}(n)$ and $e(n)$. The cost of computing $\mu\mathbf{x}(n)e(n)$ consist of $L+1$ multiplications. After computing that term, the obtention of $\mathbf{w}(n)$ requires L additions.

Then, the LMS total cost is of $2L+1$ multiplications and $2L$ additions. As the total number of operations is proportional to L, the algorithm is $O(L)$.

4.1.3 LMS with Complex Signals

In the previous sections we obtained the LMS algorithm assuming that the signals $d(n)$ and $\mathbf{x}(n)$ were real. In a great number of applications, the signals involved can be modeled as complex quantities. For this reason, it is necessary to have a proper formulation of the LMS (or any adaptive algorithm) for the case of complex signals. For complex signals $d(n)$ and $\mathbf{x}(n)$ the correlation matrix and cross correlation vector can be defined as

$$\mathbf{R_x} = E\left[\mathbf{x}(n)\mathbf{x}^H(n)\right], \quad \mathbf{r}_{\mathbf{x}d} = E\left[d^*(n)\mathbf{x}(n)\right], \tag{4.10}$$

where $(\cdot)^*$ and $(\cdot)^H$ denote conjugation and conjugated transposition operations. The optimal Wiener solution (which would be a complex vector) that minimizes $E\left[|e(n)|^2\right]$ has the same mathematical form as for the real case, but using the expressions in (4.10). Following the same approach as in Sect. 4.1.1 we obtain that the LMS for complex signals can be written as

$$\mathbf{w}(n) = \mathbf{w}(n-1) + \mu\mathbf{x}(n)e^*(n), \quad \mathbf{w}(-1). \tag{4.11}$$

In the following, we will continue considering that the signals are real. The corresponding extensions for complex signals are straightforward and the reader can obtain them easily.

4.2 NLMS Algorithm

It turns out that the LMS update at time n is a scaled version of the regression vector $\mathbf{x}(n)$, so the "size" of the update in the filter estimate is therefore proportional to the norm of $\mathbf{x}(n)$. Such behavior can have an adverse effect on the performance of LMS in some applications, e.g., when dealing with speech signals, where intervals of speech activity are often accompanied by intervals of silence. Thus, the norm of the regression vector can fluctuate appreciably. This issue can be solved by normalizing the update by $\|\mathbf{x}(n)\|^2$, leading to the *Normalized Least Mean Square* (NLMS) algorithm. However, this algorithm might be derived in different ways, leading to interesting interpretations on its operation mode.

4.2.1 Approximation to a Variable Step Size SD Algorithm

Consider the SD recursion (4.1) but now with a time varying step size $\mu(n)$. The idea is to find the step size sequence that achieves the maximum speed of convergence. Particularly, at each time step the value $\mu(n)$ should be chosen to minimize the MSE evaluated at $\mathbf{w}(n)$. From (3.14) and (2.25), the EMSE can be expressed as

$$\xi(n) = \tilde{\mathbf{w}}^T(n)\mathbf{R}_\mathbf{x}\tilde{\mathbf{w}}(n). \tag{4.12}$$

Equation (4.12) is quadratic in μ, and its solution leads to the sequence

$$\mu^o(n) = \frac{\tilde{\mathbf{w}}^T(n-1)\mathbf{R}_\mathbf{x}^2\tilde{\mathbf{w}}(n-1)}{\tilde{\mathbf{w}}^T(n-1)\mathbf{R}_\mathbf{x}^3\tilde{\mathbf{w}}(n-1)}. \tag{4.13}$$

Now, in order to get the stochastic gradient approximation algorithm, the true statistics should be replaced using (4.2). The resulting update corresponds to the NLMS, and has the form

$$\mathbf{w}(n) = \mathbf{w}(n-1) + \frac{\mathbf{x}(n)}{\|\mathbf{x}(n)\|^2}e(n). \tag{4.14}$$

So far, the NLMS can be viewed as a stochastic approximation to the SD with maximum speed of convergence in the EMSE. Given the link between SD and LMS algorithm, it might be possible to derive the NLMS from an LMS. Actually, (4.14) can be interpreted as an LMS with a time dependent step size[2]

$$\mu(n) = \frac{1}{\|\mathbf{x}(n)\|^2}. \tag{4.15}$$

From all the possible sequences $\mu(n)$ that can be applied to the LMS, the question is whether the one in (4.15) is optimal in another sense. The answer is yes, and the idea is to find $\mu(n)$ in an LMS to minimize the squared value of the a posteriori output estimation error. The a posteriori output estimation error is given by:

$$e_p(n) = d(n) - \mathbf{w}^T(n)\mathbf{x}(n), \tag{4.16}$$

that is, the estimation error computed with the updated filter. From (4.3), and using a time dependent step size, it can be obtained:

$$|e_p(n)|^2 = \left[1 - \mu(n)\|\mathbf{x}(n)\|^2\right]^2 |e(n)|^2.$$

Then, it is easy to show that this expression is minimized when $\mu(n)$ is chosen according to (4.15). Actually, in this case the a posteriori error is zero.

Consider the scenario where the desired signal follows the linear regression model (2.11). Then, since the additive noise $v(n)$ is present in the environment, by zeroing the a posteriori error the adaptive filter is forced to compensate for the effect of a noise signal which is in general uncorrelated with the adaptive filter input signal. This will lead to a high misadjustment.[3] For this reason, an additional step size μ is included in the NLMS to control its final error, giving the recursion

$$\mathbf{w}(n) = \mathbf{w}(n-1) + \frac{\mu}{\|\mathbf{x}(n)\|^2 + \delta}\mathbf{x}(n)e(n). \tag{4.17}$$

The value of δ is included in order to avoid the numerical difficulties of dividing by a very small number. In general, it is a small positive constant commonly referred as the *regularization parameter*.

4.2.2 Approximation to an NR Algorithm

We have previously introduced the regularization parameter for the Newton-Raphson algorithm. From (3.22), the regularized recursion of the NR algorithm has the form

---

[2] In the marginal case $\|\mathbf{x}(n)\| = 0$, the direction of update will be the null vector so there is no need to compute $\mu(n)$.

[3] The misadjustment will be properly defined in Sect. 4.5 where we analyze the convergence of adaptive filters. For now, it can be seen as the ratio between the steady state EMSE and the MMSE.

$$\mathbf{w}(n) = \mathbf{w}(n-1) + \mu \left(\delta \mathbf{I}_L + \mathbf{R_x}\right)^{-1} \left[\mathbf{r_{xd}} - \mathbf{R_x}\mathbf{w}(n-1)\right]. \qquad (4.18)$$

The next step is to replace the second order statistics by the instantaneous estimates given by (4.2), which leads to

$$\mathbf{w}(n) = \mathbf{w}(n-1) + \mu \left[\delta \mathbf{I}_L + \mathbf{x}(n)\mathbf{x}^T(n)\right]^{-1} \mathbf{x}(n) \left[d(n) - \mathbf{x}^T(n)\mathbf{w}(n-1)\right].$$
$$(4.19)$$

The matrix that needs to be inverted in this recursion changes at each iteration, so performing the inversion at each time step would be computationally intensive. However, this can be avoided by using the *matrix inversion lemma* (also known as *Woodbury matrix identity*) [2], which can be stated as:

Lemma 4.1 *[Matrix inversion lemma] Let* $\mathbf{A} \in \mathbb{R}^{P \times P}$ *and* $\mathbf{D} \in \mathbb{R}^{Q \times Q}$ *be invertible matrices, and* $\mathbf{B} \in \mathbb{R}^{P \times Q}$ *and* $\mathbf{C} \in \mathbb{R}^{Q \times P}$ *arbitrary rectangular matrices. Then, the following identity holds:*

$$(\mathbf{A} + \mathbf{BDC})^{-1} = \mathbf{A}^{-1} - \mathbf{A}^{-1}\mathbf{B} \left(\mathbf{D}^{-1} + \mathbf{CA}^{-1}\mathbf{B}\right)^{-1} \mathbf{CA}^{-1}. \qquad (4.20)$$

By setting $\mathbf{A} \equiv \delta \mathbf{I}_L$, $\mathbf{B} \equiv \mathbf{x}(n)$, $\mathbf{D} \equiv 1$ and $\mathbf{C} \equiv \mathbf{x}^T(n)$, we can write

$$\left[\delta \mathbf{I}_L + \mathbf{x}(n)\mathbf{x}^T(n)\right]^{-1} = \delta^{-1}\mathbf{I}_L - \frac{\delta^{-2}\mathbf{x}(n)\mathbf{x}^T(n)}{1 + \delta^{-1}\|\mathbf{x}(n)\|^2}, \qquad (4.21)$$

so multiplying both sides by $\mathbf{x}(n)$ gives

$$\left[\delta \mathbf{I}_L + \mathbf{x}(n)\mathbf{x}^T(n)\right]^{-1}\mathbf{x}(n) = \delta^{-1}\mathbf{x}(n) - \frac{\delta^{-2}\mathbf{x}(n)\|\mathbf{x}(n)\|^2}{1 + \delta^{-1}\|\mathbf{x}(n)\|^2} = \frac{\mathbf{x}(n)}{\delta + \|\mathbf{x}(n)\|^2}. \qquad (4.22)$$

Recognizing the last term of (4.19) as the error $e(n)$ and replacing (4.22) in it, we arrive at the NLMS recursion (4.17).

From Sect. 4.2.1 we can expect the NLMS with $\mu = 1$ to achieve the maximum speed of convergence as it is the stochastic approximation of the SD algorithm with maximum speed of convergence. Here, we showed the relation with the NR algorithm, from which we showed in Sect. 3.3 that it converges in one iteration when $\mu = 1$. Moreover, we see from (3.24) that as μ goes further away from 1 (but within the stable interval between 0 and 2), the convergence of NR becomes slower (as the magnitude of the mode of convergence gets closer to 1). However, we might wonder whether the NLMS would retain these characteristics given the fact that it uses estimates of the true statistics. Although the proper convergence analysis will be performed later, we would like to give an intuitive idea here. Lets assume again the linear regression model (2.11) and consider the recursion for the weight error vector of the NLMS (without regularization parameter to keep things simpler):

$$\tilde{\mathbf{w}}(n) = \left[\mathbf{I}_L - \mu \frac{\mathbf{x}(n)\mathbf{x}^T(n)}{\|\mathbf{x}(n)\|^2} \right] \tilde{\mathbf{w}}(n-1) - \frac{\mu}{\|\mathbf{x}(n)\|^2}\mathbf{x}(n)v(n), \qquad (4.23)$$

where $v(n) = d(n) - \mathbf{x}^T(n)\mathbf{w}_T$. The speed of convergence will be controlled by the homogeneous part of the equation. Note that at each time step, the weight error vector is multiplied by a matrix with $L-1$ eigenvalues equal to one and one eigenvalue equal to $1-\mu$. If $|1-\mu| < 1$ and the input signal excites all modes, the algorithm would be stable and the filter update will reach a steady state. If $\mu = 1$, the above mentioned matrix projects the weight error vector into the orthogonal subspace spanned by the regressor $\mathbf{x}(n)$. Then, having the nonunity eigenvalue equal to zero should increase the speed of convergence compared with the cases $\mu \neq 1$ because the associated mode should be driven to zero. This is showing the following tradeoff: an improvement in the misadjustment is tied to a decrease in the speed of convergence. Moreover, with $0 < \mu < 2$ and a constant $0 < \alpha < 1$, a step size $\mu = 1 + \alpha$ would lead to the same convergence speed as $\mu = 1 - \alpha$, but since the nonhomogeneous part of (4.23) is proportional to μ, the former will lead to a higher misadjustment. These issues will be explored later more rigourously when studying the convergence analysis.

4.2.3 NLMS as the Solution to a Projection Problem

The NLMS can also be derived as the exact solution to a projection problem. In this case, for positive scalars μ and δ, the problem can be stated as finding the estimate $\mathbf{w}(n)$ that solves:

$$\min_{\mathbf{w}(n)\in\mathbb{R}^L} \|\mathbf{w}(n) - \mathbf{w}(n-1)\|^2 \text{ subject to } e_p(n) = \left[1 - \frac{\mu\|\mathbf{x}(n)\|^2}{\delta + \|\mathbf{x}(n)\|^2} \right] e(n). \quad (4.24)$$

The constraint in this optimization problem can be shown to be an hyperplane in $\mathbb{R}^L$. As we are looking for $\mathbf{w}(n)$ as the vector onto this hyperplane closest to $\mathbf{w}(n-1)$ (so the norm of the update $\Delta\mathbf{w}(n)$ will be minimum), the solution to (4.24) can be thought as the orthogonal projection of $\mathbf{w}(n-1)$ onto the mentioned hyperplane. The equivalent form for the constraint is:

$$\mathbf{x}^T(n)\Delta\mathbf{w}(n) = -e_p(n) + e(n) = \frac{\mu\|\mathbf{x}(n)\|^2}{\delta + \|\mathbf{x}(n)\|^2}e(n), \qquad (4.25)$$

In this way the solution of (4.24) can be obtained through the minimum norm solution of (4.25). The minimum norm solution to this underdetermined system can be found by means of the Moore-Penrose pseudoinverse [1], leading to[4]:

---

[4] The notation $\mathbf{A}^\dagger$ denotes the Moore-Penrose pseudoinverse of matrix $\mathbf{A}$ (see Chap. 5 for further details). When $\mathbf{A} = \mathbf{x}^T(n)$ it can be shown that:

$$\Delta \mathbf{w}^o(n) = \left[\mathbf{x}^T(n)\right]^{\dagger} \frac{\mu \|\mathbf{x}(n)\|^2}{\delta + \|\mathbf{x}(n)\|^2} e(n) = \frac{\mu}{\delta + \|\mathbf{x}(n)\|^2} \mathbf{x}(n) e(n), \qquad (4.26)$$

which is the regularized NLMS update.

Now, the NLMS can be seen as an algorithm that at each time step computes the new estimate by doing the orthogonal projection of the old estimate onto the plane generated by $e_p(n) - \left[1 - \frac{\mu \|\mathbf{x}(n)\|^2}{\delta + \|\mathbf{x}(n)\|^2}\right] e(n) = 0$. Notice that when $\mu = 1$ and $\delta = 0$, the projection is done onto the space $e_p(n) = 0$, which agrees with the interpretation found in Sect. 4.2.1. Looking at the NLMS as a projection algorithm will be actually extended later to the family of affine projection algorithms. In Chap. 5 we will analyze more deeply the concept of orthogonal projections and their properties.

4.2.4 One More Interpretation of the NLMS

Another interesting connection can be made between NLMS and LMS algorithms. Consider the data available at time n, i.e., $d(n)$ and $\mathbf{x}(n)$, and an initial estimate $\mathbf{w}(n-1)$. In [3], it is shown that using an LMS with step size μ, starting at $\mathbf{w}(n-1)$ an iterating repeatedly *with the same input-output pairs*, the final estimate will be the same as the one obtained by performing a single NLMS update with step size equal to one. Although a proper proof is provided in [3], an intuitive explanation is provided here.

Since $d(n)$ and $\mathbf{x}(n)$ are fixed, an error surface can be associated, which depends only on the filter coefficients. If the time index is dropped (to emphasize that the input and output are fixed) this surface can be expressed as:

$$J(\mathbf{w}) = |e^2| = d^2 + \mathbf{w}^T \mathbf{x}\mathbf{x}^T \mathbf{w} - 2d\mathbf{w}^T \mathbf{x}. \qquad (4.27)$$

The LMS will perform several iterations at this surface. Using the subscript i to denote the iteration number, then

$$\mathbf{w}_i = \mathbf{w}_{i-1} + \mu \mathbf{x}(d - \mathbf{w}_{i-1}^T \mathbf{x}), \ \mathbf{w}_0 = \mathbf{w}(n-1). \qquad (4.28)$$

If μ is small enough to guarantee the stability of the algorithm, i.e., if μ is chosen so that $\mu < 2\|\mathbf{x}(n)\|^{-2}$, (4.28) can be interpreted as an SD search on the surface (4.27). In the limit, its minimum will be found. This minimum will satisfy

$$\mathbf{x}\mathbf{x}^T \mathbf{w}_{\min} = d\mathbf{x}.$$

(Footnote 4 continued)

$$\left[\mathbf{x}^T(n)\right]^{\dagger} = \frac{\mathbf{x}(n)}{\|\mathbf{x}(n)\|^2}.$$

There is an infinite number of solutions to this problem, but they can be written as

$$\mathbf{w}_{\min} = \frac{\mathbf{x}}{\|\mathbf{x}\|^2} d + \mathbf{x}^{\perp},$$

where $\mathbf{x}^{\perp}$ is any vector in the orthogonal space spanned by $\mathbf{x}(n)$. However, given the particular initial condition $\mathbf{w}_0 = \mathbf{w}(n-1)$, it is not difficult to show that

$$\mathbf{x}^{\perp} = \left[\mathbf{I}_L - \frac{\mathbf{x}\mathbf{x}^T}{\|\mathbf{x}\|^2} \right] \mathbf{w}_0.$$

Putting all together and reincorporating the time index, the final estimate from iterating repeatedly the LMS will be

$$\left[\mathbf{I}_L - \frac{\mathbf{x}(n)\mathbf{x}^T(n)}{\|\mathbf{x}(n)\|^2} \right] \mathbf{w}(n-1) + \frac{\mathbf{x}(n)}{\|\mathbf{x}(n)\|^2} d(n). \tag{4.29}$$

Rearranging the terms, (4.29) is equivalent to

$$\mathbf{w}(n-1) + \frac{\mathbf{x}(n)}{\|\mathbf{x}(n)\|^2} e(n),$$

which is the NLMS update with step size equal to one !!

This interesting interpretation provides more support to the idea of NLMS having faster convergence than LMS. At each time step, NLMS finds the minimum of (4.27) in "a single iteration", in contrast with LMS. Notice that in this case $J(\mathbf{w}_{\min}) = 0$, which is indeed equivalent to nullify the a posteriori output estimation error.

4.2.5 Computational Complexity of NLMS

Compared to the LMS, there is an extra cost in computing $\|\mathbf{x}(n)\|^2$. In general this will take L multiplications and $L - 1$ additions. An extra addition is used for the regularization, and a division is also required. The total cost is then $3L + 1$ multiplications and $3L$ additions. However, consider the case where the data has a shift structure (tapped delay line), a very common situation in practice, e.g., in channel estimation. In this case, two successive regressors, will only differ in two entries, so

$$\|\mathbf{x}(n)\|^2 = \|\mathbf{x}(n-1)\|^2 + |x(n)|^2 - |x(n-L)|^2.$$

That is, we can reuse the value of $\|\mathbf{x}(n-1)\|^2$ to compute $\|\mathbf{x}(n)\|^2$ efficiently. This means that with respect to the LMS, the NLMS requires in this case an extra computation of 4 multiplications, 2 additions and 1 division.

Fig. 4.1 Scheme of an adaptive noise canceler (ANC)

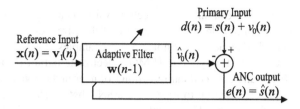

4.2.6 Example: Adaptive Noise Cancelation

The general idea of an adaptive noise canceler (ANC) is depicted in Fig. 4.1. One sensor would receive the *primary input*, that works as desired signal $d(n)$ in our adaptive filter scheme on Fig. 1.1. This is formed by a signal $s(n)$, corrupted by additive uncorrelated noise $v_0(n)$, i.e.,

$$d(n) = s(n) + v_0(n), \qquad E[s(n)v_0(n-i)] = 0 \quad \forall i. \qquad (4.30)$$

Another sensor provides the *reference input*, which will be the input $x(n)$ to the adaptive filter. This sensor receives a noise $v_1(n)$ which is uncorrelated with the signal $s(n)$, but *correlated* with $v_0(n)$, i.e.,

$$x(n) = v_1(n), \quad E[s(n)v_1(n-i)] = 0 \quad \forall i, \quad E[v_0(n)v_1(n-i)] = r(i), \quad (4.31)$$

where $r(i)$ is an unknown cross correlation for lag i. In this way, the adaptive filter will produce at its output an estimate for the error corrupting the signal on the primary input, $\hat{v}_0(n)$. This estimate can then be subtracted from the primary input, so the "error signal" is actually the estimate for the signal of interest, $\hat{s}(n)$, and so it is the output of the ANC.[5] As the error signal, this is used to adjust the coefficients of the adaptive filter. In minimizing the mean square value of the output of the ANC, it ends up being the best MSE estimate of the signal, since as the reference signal $v_1(n)$ is only correlated with the noise $v_0(n)$, the signal $s(n)$ remains essentially unaffected. This can also be interpreted as maximizing the output signal to noise ratio (SNR). The positioning of the reference sensor is important for the performance of the ANC in order to satisfy (4.31) in the best possible way.

The ANC can be used in many applications [4], as adaptive speech enhancement (listening to speech in the presence of background noise), adaptive line enhancement (detecting a periodic signal in broadband background noise), acoustic echo cancelation (coupling problems between a loudspeaker and a microphone), line echo cancelation (impedance mismatch in the analog loop of the telephone network), etc.

In this example we focus on another application which is power line interference (PLI), usually seen as a combination of sinusoids at 50/60 Hz and harmonics. When

---

[5] If this subtraction is not done properly (under the control of an adaptive filter) it might lead to an increase of the output noise power.

dealing for example with electrophysiological signals, they have in common their low amplitude levels, and because of their frequency content, they can be severely perturbed by power line noise. Some examples of such signals are electrocardiogram (ECG), electromyogram (EMG), electrooculogram (EOG), electroencephalogram (EEG) and extracellular recordings of neuronal activity. Removing this interference might be essential to improve the SNR before starting any analysis on the signal, especially if the effect of interest is nearby the noise frequency.

PLI originates from many sources, including the long wires between the subject and the amplifier, the separation between the measurement points (electrodes), capacitive coupling between the subject (a volume conductor) and power lines, and the low amplitude of the desired signals [5].

A common practice to remove PLI is to use a single fixed notch filter centered at the nominal line frequency. If we use a second order FIR filter, the resulting bandwidth might be too large, leading to signal distortion and possibly the elimination of important content (for example, for diagnostic purpose). With a second order IIR filter, we can control the bandwidth with the location of the poles. However, line frequency, amplitude and phase might not be constant. If 5% variations in the frequency of the power supply can take place, a wide enough notch filter would be required to remove all that frequency band, so signal distortion is again an issue. Conversely, the center of a very narrow notch filter may not fall exactly over the interference, failing to remove it. Adaptive filters using a reference from the power outlet can track the noise fluctuations, becoming perfect candidates for this application. As in PLI the interference is sinusoidal, the noises in the ANC can be modeled as:

$$v_0(n) = A_0 \cos(\omega_0 n + \phi_0) \qquad v_1(n) = A \cos(\omega_0 n + \phi). \qquad (4.32)$$

where the phases ϕ_0 and ϕ are fixed random variables distributed uniformly in $[0, 2\pi)$. If all signals are sampled at f_s Hz, the normalized angular frequency ω_0 is equal to $2\pi f_0/f_s$, with f_0 being the frequency of the sinusoid in Hz. As the frequency of the reference signal would presumably be the same as the one from the noise $v_0(n)$ in the primary input, the objective of the adaptive filter is to adjust its amplitude and phase in order to match the noise $v_0(n)$. To do this, it needs at least two coefficients. However, in practice, the chance of having other forms of uncorrelated noise (not necessarily sinusoidal) in the primary and reference inputs, would require the use of further coefficients.

In [6] it was shown that the transfer function of the ANC, from $d(n)$ to $e(n)$, using an LMS algorithm can be approximated by

$$H(z) = \frac{E(z)}{D(z)} = \frac{z^2 - 2z \cos(\omega_0) + 1}{z^2 - 2(1-\alpha)z \cos(\omega_0) + (1-2\alpha)}, \quad \text{with } \alpha = \frac{\mu L A^2}{4}.$$
$$\tag{4.33}$$

The zeros are located on the unit circle at $z = e^{\pm j\omega_0}$. If $\alpha \ll 1$, then

$$(1-\alpha)^2 = 1 - 2\alpha + \alpha^2 \approx 1 - 2\alpha.$$

So the poles are approximately located at $z = (1 - \alpha)e^{\pm j\omega_0}$. This shows that the ANC behaves as a second order IIR filter with central frequency ω_0. Therefore, if ω_0 is time variant (to include in the model the potential slow drift of the PLI), the adaptive filter has the potential to track these changes. Moreover, the step size μ controls the location of the poles. Since the sharpness of the notch is determined by the closeness between the zeros and the poles of $H(z)$, decreasing μ leads to a sharper notch filter. Particularly, the 3 dB bandwidth can be approximated by

$$Bw_{\text{LMS}} \approx 2\alpha = \frac{\mu L A^2}{2} \text{ radians} \qquad \left(\text{or } \frac{\mu L A^2 f_s}{4\pi} \text{ Hz}\right). \tag{4.34}$$

To analyze the differences that might arise when using an NLMS instead of the LMS algorithm, remember that the NLMS can be seen as a variable step size LMS, i.e.,

$$\mu_{\text{LMS}}(n) = \frac{\mu_{\text{NLMS}}}{\|\mathbf{x}(n)\|^2}. \tag{4.35}$$

However, for large adaptive filters $\|\mathbf{x}(n)\|^2 \approx L\sigma_x^2$, where σ_x^2 is the power of the input process. Since in this case the input to the filter is a sinusoid with an uniformly distributed random phase, $\sigma_x^2 = A^2/2$. Replacing all this in (4.34), we see that the bandwidth with the NLMS algorithm depends only on the step size μ, so

$$Bw_{\text{NLMS}} \approx \mu_{\text{NLMS}} \text{ radians} \qquad \left(\text{or } \frac{\mu_{\text{NLMS}} f_s}{2\pi} \text{ Hz}\right). \tag{4.36}$$

To test the ANC numerically, we use a recording of extracellular neural activity (sampled at 28 kHz) acquired from a depth electrode implanted in the medial temporal lobe of an epileptic patient [7]. The high impedance electrodes used for this type of recordings ($\approx 500 \, k\Omega$) would act like antennas, picking up PLI from nearby electrical equipments (notice that in a hospital environment there are many pieces of electrical equipment, and switching them off might not be an option). In this application, when possible, the reference signal can be taken from a wall outlet with proper attenuation.

We focus on the so called gamma band, which has a frequency span from 30 to 100 Hz. The smaller amplitude of this signal compared to the ones associated to slower oscillations makes it even more prone to end up masked by the PLI. As we will see later, the power of the PLI in this recording was at least 15 dB above the power of the signal. However, it has been found that gamma oscillations are related to several neuropsychiatric disorders (schizophrenia, Alzheimer's Disease, epilepsy, ADHD, etc.) and memory processing [8, 9]. Therefore, this application requires a notch filter with small rejection bandwidth but with the ability to track the possibly time varying PLI.

For comparison purposes, we also filtered the recorded signal with second order IIR notch filters with central frequencies f_0 of 60 and 61 Hz. These filters were designed to have a 3 dB bandwidth close to 1 Hz. To achieve the same bandwidth with the NLMS algorithm we used (4.36), leading to $\mu = 0.000225$. We also use a regularization constant $\delta = 1 \times 10^{-7}$. We also implemented an LMS algorithm with the same step size as in the NLMS. Both adaptive filters used $L = 32$ coefficients.

On the top right panel of Fig. 4.2 we see an extract of each signal during the transient period of the adaptation. The original gamma signal is clearly dominated by the 60 Hz PLI. The adaptive filters follow the perturbed signal rather closely as the filter coefficients have not reached yet the appropriate values. The adaptive filters reached steady state performance in less than 4 s, although they are kept running continuously. The bottom right panel is analogous to the previous one but once the steady state has been reached. The output of the NLMS based ANC is almost indistinguishable from the one of the fixed IIR notch filter with $f_0 = 60$ Hz, exposing the frequency content of the actual neuronal signal above and below 60 Hz. The output using the LMS filter is close to the one of the NLMS, but some differences appear (e.g., near $t = 15$ ms or $t = 70$ ms). To gain further insight on the operation of the filters, the left panel shows the power spectrum of each signal during steady state performance. They were constructed by averaging 300 periodograms with a resolution close to 0.5 Hz. The PLI is over 15 dB above the gamma signal, which explains why the recorded gamma signal in the right panel is dominated by the 60 Hz sinusoid. If a fixed IIR notch filter with $f_0 = 61$ Hz and $Bw = 1$ Hz is used, it completely fails to reject the PLI. It should be noticed that the difference with the actual PLI frequency is less than 2%, which exposes the previously mentioned issues on using a very narrow fixed IIR notch. We also confirm that the signal obtained using the NLMS algorithm is almost the same as the one with the fixed IIR notch centered at $f_0 = 60$ Hz. On the other hand, the LMS filter exhibits attenuation of the signal below 60 Hz and certain distortion above 60 Hz, which explains the differences previously seen in the bottom right panel.

Although in this example the frequency of the PLI was rather stationary, the adaptive filter has the potential to track its variations. However, as we mention later in this chapter, the tracking performance improves in general with larger step sizes. Therefore, there is a tradeoff between good tracking performance and fast convergence (large μ) and small misadjustment and rejection bandwidth (small μ). In this context, it sounds very appealing to explore the use of variable step size techniques [10]. In addition, the PLI can be modeled to digitally generate the reference when it is difficult to simultaneously record it with the primary input, due for example, to hardware limitations [11].

4.3 Leaky LMS Algorithm

The LMS algorithm can suffer from a potential instability problem when the regression data is not sufficiently exciting. This is generally the case when the covariance matrix of the input regressors is singular or close to singular. When this happens, the weight estimates can drift and grow unbounded. The need for a solution to this problem becomes more important when the data is quantized, where finite precision errors can favor the drift of the filter coefficients even in the case of zero noise.

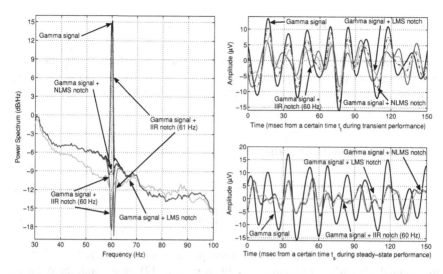

Fig. 4.2 Performance of different ANCs. The fixed IIR and the NLMS notch filters were designed to have a bandwidth of 1 Hz. The step size for the LMS algorithm was chosen to be the same as the one of the NLMS. The spectrum on the left panel were calculated after the adaptive filters have reached steady state performance

For any $\beta > 0$, a modification introduced to the cost function (4.5) leads to:

$$\hat{J}\left(\mathbf{w}(n-1)\right) = |e(n)|^2 + \beta \|\mathbf{w}(n-1)\|^2, \tag{4.37}$$

with the *Leaky LMS algorithm* as the result of the optimization problem, i.e.,

$$\mathbf{w}(n) = (1 - \mu\beta)\,\mathbf{w}(n-1) + \mu\mathbf{x}(n)e(n), \quad \mathbf{w}(-1). \tag{4.38}$$

When the input correlation matrix has a zero (or close to zero) eigenvalue, its associated mode will not converge and might actually become unstable. The leak factor will be added to the eigenvalues of the input correlation matrix. The effect would be equivalent to the addition of some white background noise to the input regressors (*dithering*). Nevertheless, we will see later that this solution for the drifting problem comes at the expense of introducing a bias to the optimal MSE solution. The computational complexity of the leaky LMS is basically the same of the LMS with the exception of L extra multiplications needed to compute the term $(1 - \mu\beta)\,\mathbf{w}(n-1)$.

4.4 Sign Algorithms

Some alternative LMS based algorithms have the feature of reducing computational cost. In doing so, they use a quantized version of the error and/or input signals. We focus here on a popular family of algorithms using the sign function as the quantizer.

4.4.1 Sign Error Algorithm

Instead of using the MSE cost function, consider the mean absolute error (MAE) criterion

$$J(\mathbf{w}) = E[|e(n)|] = E\left[|d(n) - \mathbf{w}^T\mathbf{x}(n)|\right]. \tag{4.39}$$

This is a convex function. Under certain assumptions on the processes $d(n)$ and $x(n)$, the optimal filter $\mathbf{w}^*$ that minimizes (4.39) is unique [12]. In general, the optimal filter will be different from the one associated with the MSE criterion, i.e., the Wiener filter. However, in several cases of interest, both solutions will be very close to each other.

In order to minimize (4.39), an SD method can be used. The gradient of the cost function should be computed, leading to

$$\nabla_{\mathbf{w}} J\left(\mathbf{w}(n-1)\right) = -E\left\{\text{sign}\left[e(n)\right]\mathbf{x}(n)\right\}, \tag{4.40}$$

where sign[·] is the sign function. Then, the iterative method would be

$$\mathbf{w}(n) = \mathbf{w}(n-1) + \mu E\left\{\text{sign}\left[e(n)\right]\mathbf{x}(n)\right\}.$$

To find a stochastic gradient approximation, the same ideas used for the LMS can be applied. This means that the gradient can be approximated by dropping the expectation in (4.40), or using the gradient of the cost function generated from approximating (4.39) by the (instantaneous) absolute value of the error. In any case, the result is the *Sign Error algorithm* (SEA):

$$\mathbf{w}(n) = \mathbf{w}(n-1) + \mu\mathbf{x}(n)\text{sign}\left[e(n)\right], \quad \mathbf{w}(-1). \tag{4.41}$$

The operation mode of this algorithm is rather simple. If $\hat{d}(n) = \mathbf{x}^T(n)\mathbf{w}(n-1)$ is underestimating (overestimating) $d(n)$, the error will be positive (negative), so the filter is updated in a way that $\mathbf{x}^T(n)\mathbf{w}(n)$ is less of an underestimate (overestimate).

One advantage of this algorithm is its low computational complexity. If the step size is selected as a power of 2^{-1}, the evaluation of the update can be implemented digitally very efficiently by means of shift registers. In this case, the number of multiplications come only from the filtering process for computing the output of the adaptive filter, i.e., L multiplications. The total number of sums remains $2L$.

Nevertheless, the simplification in computations comes at the expense of slower convergence. Actually, as it will be seen later in the simulation results on adaptive equalization, the shape of the learning curve is quite peculiar. The initial convergence of the MSE is very slow and later there is a quick drop. To understand this effect, the SEA can be put as

$$\mathbf{w}(n) = \mathbf{w}(n-1) + \frac{\mu}{|e(n)|}\mathbf{x}(n)e(n).$$

This can be interpreted as an LMS with a variable step size $\mu(n) = \mu/|e(n)|$. However, this step size grows as the algorithm approaches convergence, so to guarantee stability, a very small μ should be used. On the other hand, at the beginning of the adaptation process the error will be large, so $\mu(n)$ will be very small, and the algorithm will be very slow. After the error decreases sufficiently, the step size grows enough to speed up convergence.

There is another important feature of the SEA. Consider the linear regression model (2.11). Assume that the power of the noise is not too large and an adaptive filter is close to its steady state so the error is small. Also, assume that at time n this error is (slightly) positive. In this case, the updates of the LMS and SEA will be very similar (with step sizes properly chosen). However, if a very large (positive) sample of noise $v(n)$ would have appeared (e.g., impulsive noise or double talk situation in echo cancelation), the outcome would have been very different. The update of the SEA would have been exactly the same, since it relies only on the (unchanged) sign of the error, so the filter estimate would have not been perturbed. In contrast, the LMS will suffer a large change due to the resulting large error sample. The robust performance against perturbations is a distinctive feature of the SEA. In Sect. 6.4 we will elaborate more on this important feature.

4.4.2 Sign Data Algorithm

With the same motivation of reducing the computational cost of the LMS, the sign function can be applied to the regressor. The resulting *Sign Data algorithm* (SDA), also called Sign Regressor algorithm, has the recursion

$$\mathbf{w}(n) = \mathbf{w}(n-1) + \mu\text{sign}\,[\mathbf{x}(n)]\,e(n),\ \mathbf{w}(-1), \tag{4.42}$$

where the sign function is applied to each element in the regressor. It should be clear that the computational complexity of SDA is the same as the one of the SEA.

It should also be noticed that there are only 2^L possible directions for the update as the result of quantizing the regressor. On average, the update might not be a good approximation to the SD direction. This will have an impact on the stability of the algorithm. Although the SDA is stable for Gaussian inputs, it would be easy to find certain inputs that result in convergence for the LMS but not for the SDA [13].

On the other hand, the SDA does not suffer from the slow convergence of the SEA. Similarly to what was previously done for the SEA, the SDA can also be interpreted in terms of the LMS in the following way:

$$\mathbf{w}(n) = \mathbf{w}(n-1) + \mathbf{M}(n)\mathbf{x}(n)e(n),$$

where $\mathbf{M}(n)$ is a diagonal matrix with its i-th entry being $\mu_i(n) = \mu/|x(n-i)|$. This means that each coefficient of the filter has its own step size. Although this is also a time varying step size as in the SEA, its dynamics are independent on the filter convergence, in contrast with the SEA. Actually, the convergence of SDA and LMS can be very similar [14].

4.4.3 Example: Adaptive Equalization

Channel equalization is a very important topic in digital communications [15, 16]. In band limited communications systems, a particular important impairment is the distortion introduced by the channel (besides the additive noise in it). This is reflected in the phenomenon known as *intersymbol interference* (ISI). As more input symbols are introduced in the channel per unit of time, the output pulses (each corresponding to an input symbol) begin to overlap considerably. In this way, the input symbols begin to interfere with each other and the receiver is not able to distinguish the individual pulses corresponding to each of the input symbols. The reason for this phenomenon is the limited bandwidth of real channels, and their departure from an ideal flat frequency response [15].

In precise terms, we can write the action of a communication channel over a sequence of input symbols (after a proper sampling procedure) as

$$y(n) = \sum_{i=0}^{M} x(n-i)h_i + v(n)$$

$$= h_0 x(n) + \underbrace{\sum_{i=1}^{M} x(n-i)h_i}_{\text{ISI}} + v(n), \qquad (4.43)$$

where $\mathbf{h} = [h_0, h_1, \ldots, h_M]$ represents the finite length impulse response of the channel, and $v(n)$ is the additive noise introduced to the channel. Assuming that the desired symbol by the receiver at time n is $x(n)$, it is clear that the effect of the ISI term given by $\sum_{i=1}^{M} x(n-i)h_i$ will decrease the quality of service required. The detector at the receiver will use $y(n)$ to estimate the symbol $x(n)$. Because of the ISI term, it will see an increased noise floor which will degrade its performance. It is important to note, that in contrast with the impairment caused by the noise $v(n)$, the problem of the ISI cannot be solved increasing the energy of the transmitted symbols.

The use of spectrum partitioning techniques (where the channel is divided into narrowband sections, which present very limited ISI), as Orthogonal Frequency Division Multiplexing (OFDM) in wireless communications [17], permits to alleviate considerably the problem of ISI on those applications. Here, we will consider a solution based on equalization using adaptive filtering. In applications such as single carrier wired communications, adaptive equalization techniques have been successfully applied [16]. Adaptive equalization solutions can also be applied for the reduction of ISI on high speed digital circuits [18]. The general idea of using equalization can be summarized in the choice of an appropriate impulse response $\mathbf{w} = \begin{bmatrix} w_{-L+1}, \ldots, w_0, \ldots, w_{L-1} \end{bmatrix}$ to be applied to the channel output $y(n)$,

$$\hat{x}(n) = \sum_{i=-L+1}^{L-1} w_i y(n-i), \tag{4.44}$$

in such a way that the error $e(n) = x(n) - \hat{x}(n)$ should be small in some appropriate sense. For example, we could consider obtaining $\mathbf{w}$ as the solution to[6]:

$$\mathbf{w} = \arg \min_{\mathbf{w} \in \mathbb{R}^{2L+1}} E\left[\left|x(n) - \sum_{i=-L+1}^{L-1} w_i y(n-i)\right|^2\right] \tag{4.45}$$

Notice the fact that the filtering is done over L samples before and after $y(n)$. This is a consequence of the existence of a natural delay, introduced by the channel $\mathbf{h}$, which affects the transmitted symbols. The solution of (4.45) can be easily computed using the tools presented in Chap. 2, and require the knowledge of the statistics of $x(n)$, $v(n)$ and $\mathbf{h}$. This clearly precludes the use of the solution of (4.45) in real world applications. Therefore, an adaptive filtering solution is proposed next.

4.4.3.1 Adaptive Solution

We will refer to the system schematized in Fig. 4.3. As explained above, the adaptive filter $\mathbf{w}(n) \in \mathbb{R}^{2L+1}$ acts on the channel outputs as

$$\hat{x}(n) = \sum_{i=-L+1}^{L-1} w_i(n-1)y(n-i). \tag{4.46}$$

---

[6] This is not the only possible cost function that could be considered. Another popular solution is to obtain a filter $\mathbf{w}$ that completely inverts the channel $\mathbf{h}$, without taking into account the noise $v(n)$. This solution called *zero forcing* (ZF) [15] completely eliminates the ISI at the cost of possibly increasing the influence of the noise. However, when the noise is sufficiently small and the channel $\mathbf{h}$ does not present nulls in its frequency response, ZF offers a good performance.

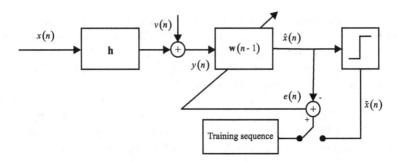

Fig. 4.3 Adaptive equalization setting

At the beginning of the adaptation, transmitter and receiver need a predefined sequence of symbols that is transmitted through the channel. This is called the *training sequence*, and in the context of Fig. 1.1 is the desired signal. At the beginning of the adaptation process, transmitter and receiver agree on this sequence. In this way, the channel output $y(n)$ and the training sequence form the input-output pairs used by the adaptive filter. The error $e(n)$ between the training sequence and the output $\hat{x}(n)$ is used for the adaptive filter control. During the period of time in which the adaptive filter uses the training sequence, it is said that the algorithm is working in the *training* mode. When the adaptive filter, achieves a proper level of equalization performance (usually measured looking at the bit error rate (BER) of the system), the transmitter begins to send actual symbols. Obviously, at the receiver, the training sequence should not be used anymore. Instead, the equalized symbols $\hat{x}(n)$ are feeded to the detector, obtaining the estimates $\tilde{x}(n)$ of the transmitted symbols $x(n)$. If the BER performance is good enough, these estimates could be used to compute the error $e(n) = \tilde{x}(n) - \hat{x}(n)$ and control the adaptive algorithm. In this way, the algorithm could be able to follow any change in the channel and adapt accordingly. When the adaptive algorithm is functioning in this way, it is said that it is working in the *decision directed* mode.[7]

We will consider an adaptive equalization problem where the SEA and SDA are used. We will analyze only the training mode of operation. In adaptive equalization applications where the computational resources are limited, sign algorithms might be an interesting choice. This could be the case in equalization problems for high speed digital circuits. In those problems, as the sampling rate can be very high, the time available to perform the filtering and update operations of an adaptive algorithm

---

[7] There are adaptive filtering variants for the equalization problem that do not require a training sequence. These are the so called blind adaptive filters [19]. Those filters do not need an exact reference or a reference at all, and can work directly with the channel outputs. There exists numerous algorithms of this kind. The most famous are the *Sato* algorithm [20] and the *Godard* algorithm [21]. The first basically works on a decision directed mode from the beginning of the adaptation process, whereas the second uses a modified cost function based solely on the amplitude of the channel outputs (no reference signal is specified). The interested reader on these types of algorithms can see [22].

could be small. This limitation calls for savings on the number of multiplications (by far, the more demanding operations). In this way and as discussed above, the SEA and SDA become natural choices.

We will consider a channel given by the following transfer function:

$$H(z) = 0.5 + 1.2z^{-1} + 1.5z^{-2} - z^{-3} + 0.5z^{-4}. \qquad (4.47)$$

The training sequence is known at the transmitter and receiver and it is a binary signal (BPSK) which could take the values 1 and -1 with equal probability. The variance of additive noise $v(n)$ is fixed at $\sigma_v^2 = 0.001$. With these choices, the SNR at the channel input is set at 30 dB. The length of the adaptive filters is fixed at $2L+1 = 15$.

In Fig. 4.4 we see the learning curves for the SEA and SDA. Although in the next sections we will give some insights on the problem of obtaining $E\left[|e(n)|^2\right]$ analytically, here we will obtain that curve using averaging. The curve is the result of the averaging of the instantaneous values $|e(n)|^2$ over 200 realizations of the equalization process (using different realizations of the noise and training sequence in each of them). With this, we can see the transient behavior and the steady state performance of both algorithms for this application. The values of the step sizes for the SEA and SDA are chosen as 0.0015 and 0.018 respectively. They were chosen to obtain the same level of steady state performance in both algorithms. As explained in Sects. 4.4.1 and 4.4.2, we see that the SDA presents a faster speed of convergence than the SEA. We also see the transient behavior of the SEA mentioned above (slow convergence speed at the beginning and a faster behavior at the end of the adaptation process). In Figs. 4.5 and 4.6 we see the symbol scatter plots corresponding to the transmission of a quadrature phase shift keying[8] (QPSK) [15] sequence of symbols, before and after equalization. Before equalization we see that for a detector it could be very hard to determine the exact symbol transmitted. On the other hand, we clearly see four clusters after equalization, each centered around the corresponding true symbol. It is clear that in this case the performance of the detector will be significantly improved.

4.5 Convergence Analysis of Adaptive Filters

Adaptive filters are stochastic systems in nature. For this reason, they are clearly different from the SD procedure presented in Chap. 3. Whereas convergence properties of SD can be analyzed using standard mathematical tools, convergence properties of adaptive filters must be analyzed using probability theory tools. This is also because adaptive filters can be thought as estimators that use input-output samples in order to obtain a vector that will be, in an appropriate sense, close to the optimal Wiener solution. Hence, the adaptive filter solution is a random variable,

---

[8] QPSK is a digital constellation composed by four symbols, which can be represented as complex quantities: $e^{j\pi/4}$, $e^{j3\pi/4}$, $e^{j5\pi/4}$ and $e^{j7\pi/4}$.

Fig. 4.4 MSE behavior of
the SEA and SDA for the
equalization problem

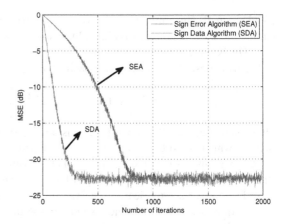

making it difficult to measure its proximity to the Wiener solution without using probability tools. Although it could be desirable to have results about almost sure convergence (that is, usual convergence for almost any realization of the stochastic processes involved [23]) to the Wiener solution, this will not be possible for adaptive algorithms with fixed step size and noisy observations.

Moreover, the issue of final performance in terms of proximity to the Wiener solution is not the only one to be considered in an adaptive filter. This is because adaptive filters are recursive estimators in nature, which leads to the issue of their stability [24–26]. This is very important, because the results about the asymptotic behavior of an adaptive filter will be valid as long as the adaptive filter is stable. The study of this issue is, in general, very difficult, basically because the adaptive filters are not only stochastic systems but also nonlinear ones. However, with certain assumptions about the input vectors some useful results can be derived. Sometimes the assumptions are very mild and reasonable [27–29] while other times they are strong and more restrictive [30, 31]. Although results with milder assumptions might be more satisfying, they are by far much more difficult to obtain. In fact, in general, during their derivation some important intuitive ideas about the functioning of the adaptive filter are obscured. On the other hand, results with stronger assumptions, although of more restrictive application, are easier to derive and highlight several important properties of the adaptive filter. In this book we will use, in general, this latter approach to show the properties of the adaptive filters.

Finally, for an adaptive filter it is important to analyze, besides its stability and steady state performance, its transient behavior. That is, how the adaptive filter evolves before it reaches its final state. As in the SD procedure, it is important how fast this is done (i.e., the speed of convergence). And even more importantly, how is the tradeoff between this speed of convergence and the steady state performance. In general, the requirements of good final performance and fast adaptation are conflicting ones. As the transient behavior analysis of adaptive filters is mathematically demanding (needing stronger assumptions), we shall not cover it in detail. However, we will provide some useful remarks about it.

Fig. 4.5 Scatter plot before equalization

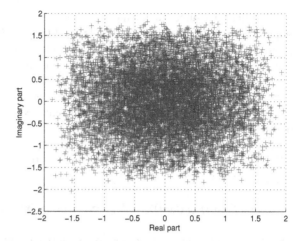

Fig. 4.6 Scatter plot after equalization

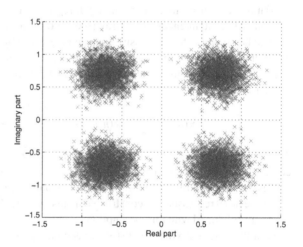

4.5.1 General Assumptions for the Convergence Analysis

We assume that the data come from the linear regression model (2.11), i.e.,

$$d(n) = y(n) + v(n) = \mathbf{w}_T^T \mathbf{x}(n) + v(n). \tag{4.48}$$

The noise $v(n)$ forms an independent and identically distributed (i.i.d.) zero mean stationary sequence, which is also independent of the zero mean input regressors $\mathbf{x}(n)$ for all n. Moreover, this noise sequence has a finite second order moment,

$$E\left[v^2(n)\right] = \sigma_v^2 < \infty. \tag{4.49}$$

We will also assume that the input regressors $\{\mathbf{x}(n)\}$ form a zero mean stationary sequence, so that for each $n \geq 0$, $E\left[\mathbf{x}(n)\mathbf{x}^T(n)\right] = \mathbf{R}_\mathbf{x}$ is a symmetric strictly positive definite $L \times L$ matrix. We will also define the variance of the scalar input $x(n)$ as $\sigma_x^2 = E\left[x^2(n)\right]$. With this assumptions on the input and the noise it is clear that $\{d(n), \mathbf{x}(n)\}$ are jointly stationary. Note also, that according to the results from Sect. 2.3, $\mathbf{w}_T$ is equal to the Wiener filter corresponding to the problem of estimating $d(n)$ given $\mathbf{x}(n)$. As the noise $v(n)$ is independent of $\mathbf{x}(n)$, it is clear that the principle of orthogonality is satisfied.

The next assumption we will make is known as the *independence assumption* [19, 32]. In precise terms it says that $\mathbf{x}(0), \mathbf{x}(1), \ldots, \mathbf{x}(n), \ldots,$ is an independent vector sequence, which means that

$$E\left[f\left(\mathbf{x}(n)\right) g\left(\mathbf{x}(m)\right)\right] = E\left[f\left(\mathbf{x}(n)\right)\right] E\left[g\left(\mathbf{x}(m)\right)\right], \quad \forall m \neq n, \qquad (4.50)$$

for arbitrary functions f and g (possibly vectorial and matricial functions of appropriate dimensions). This assumption could be true in some applications. For example, in adaptive beamforming, where the input regressor is formed by the signal picked up from different antennas in an array (see Sect. 5.7 for further details). In this case, it could be assumed that successive snapshots received by the antenna array are independent. However, the reader may guess that this assumption will not be true in the important case where the input regressors are constructed from a scalar process $x(n)$ in the following way:

$$\mathbf{x}(n) = \left[x(n)\, x(n-1) \cdots x(n-L+1)\right]^T. \qquad (4.51)$$

This vector $\mathbf{x}(n)$ is said to come from a *tapped delay line*. In this case, even if the correlation matrix $\mathbf{R}_\mathbf{x}$ is a diagonal matrix, there is a strong correlation between successive input regressors $\mathbf{x}(n)$. Given that in many typical applications of adaptive filters the input regressors are constructed in this way [4, 33], it might be expected that the results obtained with this assumption will be useless. However, it is observed that even in the case that $\mathbf{x}(n)$ is constructed according to (4.51), the theoretical results obtained using the independence assumption are, in general, in close agreement with the ones observed in typical computer simulations. This is specially true when the step size μ is small enough.

In the following we will use these assumptions indicating clearly where each one is needed. In this way, the reader will see which are the critical steps in the analysis where it is necessary to take these simplifying assumptions for obtaining closed form results. A lot of effort would be required (mathematically, numerically or both) to get those results if these assumptions were relaxed. It is also possible that we will need other assumptions for obtaining more particular results depending on what we are analyzing (stability, steady state or transient behavior). In those cases, we will mention those additional assumptions and how they will be used.

4.5.2 Stability Analysis for a Large Class of Algorithms

In this section we will analyze the stability of a large family of adaptive algorithms that encompasses several of the ones described in this chapter. The general form assumed for the algorithms is:

$$\mathbf{w}(n) = \alpha \mathbf{w}(n-1) + \boldsymbol{\mu} \mathbf{f}(\mathbf{x}(n)) e(n), \tag{4.52}$$

where α is a positive number, typically $0 < \alpha \leq 1$, $\boldsymbol{\mu} = \text{diag}(\mu_1, \mu_2, \ldots, \mu_L)$, $\mu_i > 0$, and $\mathbf{f} : \mathbb{R}^L \to \mathbb{R}^L$ is an arbitrary multidimensional function. The recursion (4.52) is sufficiently general to encompass several basic algorithms. For example, when $\alpha = 1$, $\boldsymbol{\mu} = \mu \mathbf{I}_L$ and $\mathbf{f}(\mathbf{x}(n)) = \mathbf{x}(n)$ we obtain the LMS algorithm. With the same choices for α and $\boldsymbol{\mu}$ and $\mathbf{f}(\mathbf{x}(n)) = \text{sign}[\mathbf{x}(n)]$ we obtain the SDA, and with $\mathbf{f}(\mathbf{x}(n)) = \frac{\mathbf{x}(n)}{\|\mathbf{x}(n)\|^2}$ we get the NLMS algorithm. With (4.52) we can also obtain other interesting variants as the multiple step size LMS and the Leaky LMS [19, 34].

We will obtain stability results for this general model and then specialize the results for some of the algorithms presented in this chapter. It is clear from (4.52) that the sign error algorithm discussed in 4.4.1 cannot be analyzed within this framework. This is because that algorithm presents a nonlinearity in the error, and the model (4.52) covers only algorithms that are linear in the error. For an excellent and rigorous analysis of the sign error algorithm see [12].

The stability analysis presented here will be limited to two important measures: mean stability and mean square deviation (MSD) stability. Using the misalignment vector defined in (2.23), we will say that an algorithm is *mean stable* if

$$\lim_{n \to \infty} E\left[\tilde{\mathbf{w}}(n)\right] < \infty. \tag{4.53}$$

That is, all the components of $\lim_{n \to \infty} E\left[\tilde{\mathbf{w}}(n)\right]$ are finite. In an analog manner, we will say that an algorithm is *MSD stable* when

$$\lim_{n \to \infty} E\left[\|\tilde{\mathbf{w}}(n)\|^2\right] < \infty. \tag{4.54}$$

The following are some important remarks. It is clear that in (4.53) it is included the important case in which $\mathbf{w}(n)$ is an asymptotically unbiased estimator. Although the unbiasedness property is a very useful one, in general one should not think that it is the only important property we should look at in an adaptive algorithm. It is clear that this property will be useless if the asymptotical dispersion of $\mathbf{w}(n)$ around its mean value is very large or infinite. Although a proper description of that dispersion will be given by the correlation matrix $E\left[\tilde{\mathbf{w}}(n)\tilde{\mathbf{w}}^T(n)\right]$, the quantity $E\left[\|\tilde{\mathbf{w}}(n)\|^2\right]$ will be useful too, and easier to handle. Then, the condition (4.54) permits a more insightful view on the asymptotic stability performance of an adaptive algorithm. In fact, it can be easily shown that the MSD stability implies mean stability.

We will begin our stability analysis using (4.48) into (4.52), leading to

$$\tilde{\mathbf{w}}(n) = \left[\alpha\mathbf{I}_L - \boldsymbol{\mu}\mathbf{f}\left(\mathbf{x}(n)\right)\mathbf{x}^T(n)\right]\tilde{\mathbf{w}}(n-1) - \boldsymbol{\mu}\mathbf{f}\left(\mathbf{x}(n)\right)v(n) + (1-\alpha)\mathbf{w}_T. \quad (4.55)$$

Our interest is to analyze the mean and mean square stability of $\tilde{\mathbf{w}}(n)$. For this reason, and based on the statistical properties of the input regressors, we will look for conditions on $\boldsymbol{\mu}$ that guarantee the stability of the adaptive filter. Given the fact that we restrict ourselves to the case $\boldsymbol{\mu} = \text{diag}(\mu_1, \mu_2, \ldots, \mu_L)$ with $\mu_i > 0$, which is an invertible matrix, we can work with $\tilde{\mathbf{c}}(n) = \boldsymbol{\mu}^{-1/2}\tilde{\mathbf{w}}(n)$ without loss of generality. Then, from the behavior of $\tilde{\mathbf{c}}(n)$ we can obtain precise results for the behavior of $\tilde{\mathbf{w}}(n)$, simply by taking $\boldsymbol{\mu}^{1/2}\tilde{\mathbf{c}}(n)$. Defining $\tilde{\mathbf{f}}\left(\mathbf{x}(n)\right) = \boldsymbol{\mu}^{1/2}\mathbf{f}\left(\mathbf{x}(n)\right)$, $\tilde{\mathbf{x}}(n) = \boldsymbol{\mu}^{1/2}\mathbf{x}(n)$, and $\mathbf{c}_T = \boldsymbol{\mu}^{-1/2}\mathbf{w}_T$, (4.55) can be rewritten as

$$\tilde{\mathbf{c}}(n) = \left[\alpha\mathbf{I}_L - \tilde{\mathbf{f}}\left(\mathbf{x}(n)\right)\tilde{\mathbf{x}}^T(n)\right]\tilde{\mathbf{c}}(n-1) - \tilde{\mathbf{f}}\left(\mathbf{x}(n)\right)v(n) + (1-\alpha)\mathbf{c}_T. \quad (4.56)$$

Starting from $n = 0$ we can write this equation in the following form:

$$\tilde{\mathbf{c}}(n) = \mathbf{A}(n,0)\tilde{\mathbf{c}}(-1) - \sum_{j=0}^{n}\mathbf{A}(n, j+1)\left[\tilde{\mathbf{f}}\left(\mathbf{x}(j)\right)v(j) - (1-\alpha)\mathbf{c}_T\right], \quad (4.57)$$

where the matrices $\mathbf{A}(n, i)$ are defined by

$$\mathbf{A}(n, i) = \begin{cases} \left[\alpha\mathbf{I}_L - \tilde{\mathbf{f}}\left(\mathbf{x}(n)\right)\tilde{\mathbf{x}}(n)^T\right]\ldots\left[\alpha\mathbf{I}_L - \tilde{\mathbf{f}}\left(\mathbf{x}(i)\right)\tilde{\mathbf{x}}(i)^T\right], & n \geq i \\ \mathbf{I}_L, & i = n+1 \end{cases}. \quad (4.58)$$

Equation (4.57) will be the key equation to analyze the stability of several of the algorithms previously presented. The interesting thing about this equation is that it shows the explicit dependence of $\tilde{\mathbf{c}}(n)$, or equivalently $\tilde{\mathbf{w}}(n)$, on the input regressors and the noise sequence. We will begin our analysis with the study of $E\left[\tilde{\mathbf{c}}(n)\right]$.

4.5.2.1 Mean Stability

We take expectation on both sides of (4.57) to obtain:

$$E\left[\tilde{\mathbf{c}}(n)\right] = E\left[\mathbf{A}(n,0)\right]\tilde{\mathbf{c}}(-1) - \sum_{j=0}^{n}\left\{E\left[\mathbf{A}(n, j+1)\tilde{\mathbf{f}}\left(\mathbf{x}(j)\right)v(j)\right]\right.$$
$$\left. - (1-\alpha)E\left[\mathbf{A}(n, j+1)\right]\mathbf{c}_T\right\} \quad (4.59)$$

The term $\tilde{\mathbf{c}}(-1)$ is not affected by the expectation operator because it depends on the initial condition and the true system $\mathbf{w}_T$, which are deterministic quantities. Using the independence between the noise sequence and the input, and the fact that the noise is zero mean,

$$E\left[\tilde{\mathbf{c}}(n)\right] = E\left[\mathbf{A}(n, 0)\right]\tilde{\mathbf{c}}(-1) + (1 - \alpha)\sum_{j=0}^{n} E\left[\mathbf{A}(n, j + 1)\right]\mathbf{c}_{\mathrm{T}} \qquad (4.60)$$

At this point we can use the independence assumption on the input regressors. It is then clear that

$$E\left[\mathbf{A}(n, j + 1)\right] = \prod_{i=j+1}^{n} E\left[\alpha\mathbf{I}_L - \tilde{\mathbf{f}}\left(\mathbf{x}(i)\right)\tilde{\mathbf{x}}(i)^T\right] = \mathbf{B}_{\mathbf{x}}^{n-j}, \qquad (4.61)$$

where in the first equality the order of the matrix products is the same as in (4.58). The matrix $\mathbf{B}_{\mathbf{x}} = E\left[\alpha\mathbf{I}_L - \tilde{\mathbf{f}}\left(\mathbf{x}(j)\right)\tilde{\mathbf{x}}(j)^T\right]$ is given by

$$\mathbf{B}_{\mathbf{x}} = \alpha\mathbf{I}_L - \boldsymbol{\mu}^{1/2}\mathbf{C}_{\mathbf{x}}\boldsymbol{\mu}^{1/2}, \quad \text{with } \mathbf{C}_{\mathbf{x}} = E\left[\mathbf{f}\left(\mathbf{x}(n)\right)\mathbf{x}^T(n)\right]. \qquad (4.62)$$

Therefore, we can write (4.60) as:

$$E\left[\tilde{\mathbf{c}}(n)\right] = \mathbf{B}_{\mathbf{x}}^{n+1}\tilde{\mathbf{c}}_{-1} + (1 - \alpha)\sum_{j=0}^{n} \mathbf{B}_{\mathbf{x}}^{n-j}\mathbf{c}_{\mathrm{T}}. \qquad (4.63)$$

This is a matricial equation and we will assume that $\mathbf{B}_{\mathbf{x}}$ is diagonalizable.[9] We have then the following lemma:

Lemma 4.2 *Equation (4.63) will be stable (i.e., $\lim_{n\to\infty} \| E\left[\tilde{\mathbf{c}}(n)\right] \| < \infty$), for every choice of $\tilde{\mathbf{c}}(-1)$ and $\mathbf{c}_{\mathrm{T}}$ if and only if $\mathrm{eig}_i\left[\mathbf{B}_{\mathbf{x}}\right] < 1, \ i = 1, \ldots, L$.*[10]

Proof As $\mathbf{B}_{\mathbf{x}}$ is diagonalizable, it exists an invertible matrix $\mathbf{P}$ such that $\mathbf{B}_{\mathbf{x}} = \mathbf{P}\boldsymbol{\Lambda}\mathbf{P}^{-1}$, where $\boldsymbol{\Lambda} = \mathrm{diag}\left(\lambda_1, \ldots, \lambda_L\right)$, and λ_i denotes the eigenvalues of $\mathbf{B}_{\mathbf{x}}$. It is easy to show that $\mathbf{B}_{\mathbf{x}}^n = \mathbf{P}\boldsymbol{\Lambda}^n\mathbf{P}^{-1}$. Now, we can write equation (4.63) as:

$$E\left[\tilde{\mathbf{c}}(n)\right] = \mathbf{P}\boldsymbol{\Lambda}^{n+1}\mathbf{P}^{-1}\tilde{\mathbf{c}}(-1) + (1 - \alpha)\sum_{j=0}^{n} \mathbf{P}\boldsymbol{\Lambda}^{n-j}\mathbf{P}^{-1}\mathbf{c}_{\mathrm{T}}$$

$$= \mathbf{P}\boldsymbol{\Lambda}^{n+1}\mathbf{P}^{-1}\tilde{\mathbf{c}}(-1) + (1 - \alpha)\mathbf{P}\left(\sum_{j=0}^{n} \boldsymbol{\Lambda}^{n-j}\right)\mathbf{P}^{-1}\mathbf{c}_{\mathrm{T}}$$

$$= \mathbf{P}\boldsymbol{\Lambda}^{n+1}\mathbf{P}^{-1}\tilde{\mathbf{c}}(-1) + (1 - \alpha)\mathbf{P}\left(\sum_{j=0}^{n} \boldsymbol{\Lambda}^{j}\right)\mathbf{P}^{-1}\mathbf{c}_{\mathrm{T}}. \qquad (4.64)$$

---

[9] If $\mathbf{B}_{\mathbf{x}}$ is not diagonalizable, we can always find a Jordan decomposition [35] for it, and the result from Lemma 4.2 is still valid [24].

[10] We use $\mathrm{eig}_i\left[\mathbf{A}\right]$ to denote the i-th eigenvalue of matrix $\mathbf{A}$.

If $\lambda_i < 1$, $i = 1, \ldots, L$ it is clear that $\mathbf{P}\boldsymbol{\Lambda}^{n+1}\mathbf{P}^{-1}\tilde{\mathbf{c}}(-1)$ goes to zero as $n \to \infty$. At the same time, each of the diagonal entries in $\sum_{j=0}^{n} \boldsymbol{\Lambda}^j$ converges to $\frac{1}{1-\lambda_i}$. This means that every component in $\lim_{n\to\infty} E\left[\tilde{\mathbf{c}}(n)\right]$ is finite. On the other hand, if for some i $\lambda_i \geq 1$, we have that both terms in the right hand side of (4.64) are unbounded, which means that $\lim_{n\to\infty} \|E\left[\tilde{\mathbf{c}}(n)\right]\| = \infty$. $\square$

Using (4.62) we can set the necessary and sufficient condition for the mean stability of $\tilde{\mathbf{w}}(n)$ as:

$$\left|\alpha - \text{eig}_i\left[\boldsymbol{\mu}^{1/2}\mathbf{C_x}\boldsymbol{\mu}^{1/2}\right]\right| < 1, \quad i = 1, 2, \ldots, L. \tag{4.65}$$

As we will see, in several cases of interest the matrix $\mathbf{C_x}$ will be positive definite, which implies that $\text{eig}_i\left[\boldsymbol{\mu}^{1/2}\mathbf{C_x}\boldsymbol{\mu}^{1/2}\right] > 0$, $i = 1, \ldots, L$. In these cases, the mean stability condition can be written as:

$$0 < \text{eig}_i\left[\boldsymbol{\mu}^{1/2}\mathbf{C_x}\boldsymbol{\mu}^{1/2}\right] < 1 + \alpha, \quad i = 1, 2, \ldots, L. \tag{4.66}$$

It is clear from (4.66) that a careful choice of $\boldsymbol{\mu}$ can guarantee the mean stability.

From the above proof we can see that the limiting value of $E\left[\tilde{\mathbf{c}}(n)\right]$ can be put as

$$\lim_{n\to\infty} E\left[\tilde{\mathbf{c}}(n)\right] = (1-\alpha)\mathbf{P}\left(\mathbf{I}_L - \boldsymbol{\Lambda}\right)^{-1}\mathbf{P}^{-1}\mathbf{c}_T, \tag{4.67}$$

or equivalently

$$\lim_{n\to\infty} E\left[\tilde{\mathbf{w}}(n)\right] = (1-\alpha)\boldsymbol{\mu}^{1/2}\mathbf{P}\left(\mathbf{I}_L - \boldsymbol{\Lambda}\right)^{-1}\mathbf{P}^{-1}\boldsymbol{\mu}^{-1/2}\mathbf{w}_T. \tag{4.68}$$

Evidently, if $\alpha \neq 1$ the adaptive filter will give a biased estimation when $n \to \infty$. This is the case of the Leaky LMS, as previously discussed.

In the following we particularize these results for some of the algorithms presented in the previous sections:

- **LMS algorithm**: For the LMS we have $\alpha = 1$, $\boldsymbol{\mu} = \mu\mathbf{I}_L$ and $\mathbf{f}\left(\mathbf{x}(n)\right) = \mathbf{x}(n)$. It is easy to see that condition (4.66) can be expressed as:

$$0 < \mu \, \text{eig}_i\left[\mathbf{R_x}\right] < 2, \quad i = 1, 2, \ldots, L, \tag{4.69}$$

which leads to

$$0 < \mu < \frac{2}{\text{eig}_{\max}\left[\mathbf{R_x}\right]}. \tag{4.70}$$

This is the same as (3.13), so the mean stability condition for the LMS is the same as the one for the stability of the SD algorithm. In fact the convergence dynamics of the mean weight error vector for the LMS algorithm (under the independence assumption) is the same as the one of the SD algorithm.

- **NLMS algorithm**: In this case we have $\mathbf{f}(\mathbf{x}(n)) = \frac{\mathbf{x}(n)}{\|\mathbf{x}(n)\|^2 + \delta}$, $\boldsymbol{\mu} = \mu \mathbf{I}_L$, and $\alpha = 1$. Replacing everything in (4.66), the step size μ needs to satisfy

$$0 < \mu < \frac{2}{\text{eig}_i [\mathbf{K_x}]}, \quad i = 1, \ldots, L, \tag{4.71}$$

where

$$\mathbf{K_x} = E\left[\frac{\mathbf{x}(n)\mathbf{x}^T(n)}{\|\mathbf{x}(n)\|^2 + \delta}\right]. \tag{4.72}$$

The condition (4.71) can be compactly written as:

$$0 < \mu < \frac{2}{\text{eig}_{\max} [\mathbf{K_x}]}. \tag{4.73}$$

As in the case of the LMS, if (4.73) is satisfied, the NLMS will be an asymptotically unbiased estimator of the optimal Wiener filter, as it can be seen from (4.68). The exact calculation of $\text{eig}_{\max} [\mathbf{K_x}]$ requires knowledge of the input vector probability density function (pdf). However, it is possible to bound it in a useful and appropriate way. We can write the following, given the fact that $\mathbf{K_x}$ is a symmetric positive definite matrix:

$$\text{eig}_{\max} [\mathbf{K_x}] = \max_{\mathbf{a} \in \mathbb{R}^L : \|\mathbf{a}\| = 1} \mathbf{a}^T \mathbf{K_x} \mathbf{a} \leq E\left[\max_{\mathbf{a} \in \mathbb{R}^L : \|\mathbf{a}\| = 1} \frac{|\mathbf{a}^T \mathbf{x}(n)|^2}{\|\mathbf{x}(n)\|^2 + \delta}\right] < 1, \tag{4.74}$$

which means that we can modify condition (4.73) by the following more restrictive one:

$$0 < \mu < 2. \tag{4.75}$$

Although condition (4.75) is only sufficient for the mean stability, whereas condition (4.73) is necessary and sufficient, it is by far more useful. Notice that it does not depend on any statistical information about the input vector. In addition, condition (4.75) is valid when $\delta = 0$.

- **Sign Data algorithm**: For the sign data algorithm, we have $\mathbf{f}(\mathbf{x}(n)) = \text{sign}[\mathbf{x}(n)]$, $\boldsymbol{\mu} = \mu \mathbf{I}_L$, and $\alpha = 1$. If we define

$$\mathbf{L_x} = E\left\{\text{sign}[\mathbf{x}(n)]\mathbf{x}^T(n)\right\}, \tag{4.76}$$

and assume that this matrix is positive definite, (4.66) can be compactly put as:

$$0 < \mu < \frac{2}{\text{eig}_{\max} [\mathbf{L_x}]}. \tag{4.77}$$

In this case, the value of μ that guarantees the stability in the mean depends on the statistical properties of the input through the matrix $\mathbf{L_x}$. In general, it is not possible to obtain closed form solutions for this matrix. An important case which leads to a closed form result is when $\mathbf{x}(n)$ is a zero mean Gaussian vector with correlation matrix $\mathbf{R_x}$. In that case, we obtain:

$$\mathbf{L_x} = \sqrt{\frac{2}{\pi \sigma_x^2}} \mathbf{R_x}. \tag{4.78}$$

This result can be obtained with the well known *Price theorem* [36] which permits to calculate expectations of certain nonlinear functions of correlated Gaussian random variables. In this manner, the stability condition (4.77) can be expressed as:

$$0 < \mu < \frac{\sqrt{2\pi \sigma_x^2}}{\mathrm{eig_{max}}\,[\mathbf{R_x}]}. \tag{4.79}$$

4.5.2.2 MSD Stability

In order to obtain a better picture of the way in which an adaptive filter works, besides the mean behavior, we need to look into the mean square performance. To obtain results regarding the MSD stability of the algorithm we need to take squared norms on both sides of equation (4.57) to obtain:

$$\|\tilde{\mathbf{c}}(n)\|^2 = \tilde{\mathbf{c}}^T(-1)\mathbf{A}^T(n,0)\mathbf{A}(n,0)\tilde{\mathbf{c}}(-1) - 2\sum_{j=0}^{n}\tilde{\mathbf{c}}^T(-1)\mathbf{A}^T(n,0)\mathbf{A}(n,j+1)$$

$$\times \left[\tilde{\mathbf{f}}(\mathbf{x}(j))\,v(j) - (1-\alpha)\mathbf{c_T}\right] + \sum_{i=0}^{n}\sum_{j=0}^{n}\left[\tilde{\mathbf{f}}(\mathbf{x}(i))\,v(i) - (1-\alpha)\mathbf{c_T}\right]^T$$

$$\times \mathbf{A}^T(n,i+1)\mathbf{A}(n,j+1)\left[\tilde{\mathbf{f}}(\mathbf{x}(j))\,v(j) - (1-\alpha)\mathbf{c_T}\right]. \tag{4.80}$$

If we take expectations on both sides, using the independence between the input and the noise and the fact that the noise is i.i.d., we obtain:

$$E\left[\|\tilde{\mathbf{c}}(n)\|^2\right] = \tilde{\mathbf{c}}^T(-1)E\left[\mathbf{A}^T(n,0)\mathbf{A}(n,0)\right]\tilde{\mathbf{c}}(-1)$$

$$+ 2(1-\alpha)\sum_{j=0}^{n}\tilde{\mathbf{c}}^T(-1)E\left[\mathbf{A}^T(n,0)\mathbf{A}(n,j+1)\right]\mathbf{c}_T$$

$$+ \sigma_v^2\sum_{j=0}^{n}E\left[\tilde{\mathbf{f}}^T(\mathbf{x}(j))\,\mathbf{A}^T(n,j+1)\mathbf{A}(n,j+1)\tilde{\mathbf{f}}(\mathbf{x}(j))\right]$$

$$+ (1-\alpha)^2\sum_{j=0}^{n}\sum_{k=0}^{n}\mathbf{c}_T^T E\left[\mathbf{A}^T(n,k+1)\mathbf{A}(n,j+1)\right]\mathbf{c}_T. \quad (4.81)$$

We can define $\mathbf{D}(n,j) = E\left[\mathbf{A}^T(n,j)\mathbf{A}(n,j)\right]$, which is a positive definite matrix. From the definition (4.58) and the independence assumption we obtain:

$$E\left[\mathbf{A}^T(n,k)\mathbf{A}(n,j)\right] = \begin{cases} \mathbf{D}(n,k)\mathbf{B}_{\mathbf{x}}^{k-j}, & k \geq j \\ \left(\mathbf{B}_{\mathbf{x}}^T\right)^{j-k}\mathbf{D}(n,j), & j > k \end{cases}, \quad (4.82)$$

where $\mathbf{B}_{\mathbf{x}}$ is given by (4.62). With the above equation we can write the first, second, third and fourth terms in the right hand side of (4.81) as:

$$\tilde{\mathbf{c}}^T(-1)\mathbf{D}(n,0)\tilde{\mathbf{c}}(-1), \quad (4.83)$$

$$2(1-\alpha)\sum_{j=0}^{n}\tilde{\mathbf{c}}^T(-1)\left(\mathbf{B}_{\mathbf{x}}^T\right)^{j+1}\mathbf{D}(n,j+1)\mathbf{c}_T, \quad (4.84)$$

$$\sigma_v^2\sum_{j=0}^{n}\text{tr}\left[\tilde{\mathbf{F}}_{\mathbf{x}}\mathbf{D}(n,j+1)\right], \quad (4.85)$$

$$(1-\alpha)^2\sum_{j=0}^{n}\left[\sum_{k=j}^{n}\mathbf{c}_T^T\mathbf{D}(n,k+1)\mathbf{B}_{\mathbf{x}}^{k-j}\mathbf{c}_T + \sum_{k=0}^{j-1}\mathbf{c}_T^T\left(\mathbf{B}_{\mathbf{x}}^T\right)^{j-k}\mathbf{D}(n,j+1)\mathbf{c}_T\right], \quad (4.86)$$

respectively, where $\tilde{\mathbf{F}}_{\mathbf{x}} = E\left[\tilde{\mathbf{f}}(\mathbf{x}(j))\tilde{\mathbf{f}}^T(\mathbf{x}(j))\right]$, is assumed to be positive definite.[11] Given that $\mu_i > 0$, $i = 1, 2, \ldots, L$, the MSD stability of $\tilde{\mathbf{w}}(n)$ is equivalent to the MSD stability of $\tilde{\mathbf{c}}(n)$. The latter can be characterized by the following theorem:

Theorem 4.1 *[Mean square stability] A necessary and sufficient condition for* $\lim_{n\to\infty} E\left[\|\tilde{\mathbf{c}}(n)\|^2\right] < \infty$ *for every* $\tilde{\mathbf{c}}(-1)$ *and* $\mathbf{c}_T$, *is given by the satisfaction of (4.65) and*

---

[11] In (4.85) we used the fact that $\mathbf{A}(n,j+1)$ and $\tilde{\mathbf{f}}(\mathbf{x}(j))$ are independent and that for two matrices $\mathbf{A}$ and $\mathbf{B}$ of appropriate dimensions, $\text{tr}[\mathbf{AB}] = \text{tr}[\mathbf{BA}]$.

$\exists N(\mu), \gamma(\mu)$ with $N(\mu) > 0$ and $0 < \gamma(\mu) < 1$ such that

$$\text{tr}\left[\mathbf{D}(i, j+1)\right] \leq N(\mu)\left[\gamma(\mu)\right]^{i-j}, \ \forall i \geq j. \tag{4.87}$$

Although the proof of Theorem 4.1 is not difficult, it is rather long. As we are more interested in the result, we will omit it. The reader interested in the formal proof is referred to [37].

The fact that the condition given in (4.87) is necessary and sufficient allows us to restrict, without loss of generality, to study when an exponential behavior on $\text{tr}\left[\mathbf{D}(n, k+1)\right]$ can be obtained.[12] Although there are multiple ways to accomplish this, we will restrict ourselves to a particular one. By doing this, we will obtain only sufficient conditions for the mean square stability. However, this is not a serious drawback because from the design point of view having sufficient conditions is more useful than having necessary conditions for the mean and mean square stability.

Using the definition of $\mathbf{D}(n, k+1)$, (4.58), and the independence assumption, we can write

$$\begin{aligned}
\text{tr}\left[\mathbf{D}(n, j)\right] &= \text{tr}\left[\mathbf{A}^T(n-1, j)\mathbf{G_x}\mathbf{A}(n-1, j)\right] \\
&= \text{tr}\left[\mathbf{G_x}\mathbf{A}(n-1, j)\mathbf{A}^T(n-1, j)\right] \leq \text{eig}_{\max}\left[\mathbf{G_x}\right]\text{tr}\left[\mathbf{D}(n-1, j)\right],
\end{aligned} \tag{4.88}$$

where $\mathbf{G_x} = E\left[\mathbf{E}^T(n)\mathbf{E}(n)\right]$, with $\mathbf{E}(n) = \alpha\mathbf{I}_L - \tilde{\mathbf{f}}(\mathbf{x}(n))\tilde{\mathbf{x}}(n)^T$. This can be repeated several times to obtain

$$\text{tr}\left[\mathbf{D}(n, j)\right] \leq L\left\{\text{eig}_{\max}\left[\mathbf{G_x}\right]\right\}^{n-j+1}. \tag{4.89}$$

From this, and in light of the result of Theorem 4.1, we can guarantee the mean square stability by choosing μ in such a way that

$$\text{eig}_{\max}\left[\mathbf{G_x}\right] < 1. \tag{4.90}$$

The matrix $\mathbf{G_x}$ can be put in explicit form as:

$$\mathbf{G_x} = \alpha^2\mathbf{I}_L - \alpha\mu^{1/2}\mathbf{C_x}^T\mu^{1/2} - \alpha\mu^{1/2}\mathbf{C_x}\mu^{1/2} + \mu^{1/2}\mathbf{H_x}\mu^{1/2}, \tag{4.91}$$

where $\mathbf{H_x}$ is given by

$$\mathbf{H_x} = E\left[\mathbf{f}^T(\mathbf{x}(n))\,\mu\mathbf{f}(\mathbf{x}(n))\,\mathbf{x}(n)\mathbf{x}^T(n)\right]. \tag{4.92}$$

---

[12] The fact that $N(\mu)$ depends on μ is not relevant from the point of view of the stability, because it has no influence on the asymptotic behavior of $\text{tr}\left[\mathbf{D}(n, k+1)\right]$.

As (4.90) is equivalent to asking for[13]

$$\mathbf{G_x} < \mathbf{I}_L \tag{4.93}$$

we will need to look at μ in order to guarantee this.

We will see that, although the exponential bound (4.89) might not be the tightest one, we can find with this approach some well known results on the stability of classical adaptive filters, which are actually very tight.[14]

In the following we particularize these results for some of the algorithm presented in the previous sections:

- **LMS algorithm**: Remember that in this case $\mathbf{f}(\mathbf{x}(n)) = \mathbf{x}(n)$, $\alpha = 1$, and $\boldsymbol{\mu} = \mu \mathbf{I}_L$. We need to consider (4.91) and check when $\mathbf{G_x} < \mathbf{I}_L$. We can show that in the LMS case, the matrix $\mathbf{G_x}$ can be written as:

$$\mathbf{G_x} = \mathbf{I}_L - 2\mu \mathbf{R_x} + \mu^2 \mathbf{S_x}, \tag{4.94}$$

where the matrix $\mathbf{S_x}$ is defined as:

$$\mathbf{S_x} = E\left[\|\mathbf{x}(n)\|^2 \mathbf{x}(n)\mathbf{x}^T(n)\right]. \tag{4.95}$$

The matrix $\mathbf{S_x}$ depends on the fourth order moments of the input signal which are assumed to exist. The stability condition can be found in the following lemma:

Lemma 4.3 *The matrix* $\mathbf{G_x}$ *in (4.94) will satisfy* $\mathrm{eig}_{\max}[\mathbf{G_x}] < 1$ *if and only if*

$$0 < \mu < \frac{2}{\mathrm{eig}_{\max}\left[\mathbf{R_x}^{-1}\mathbf{S_x}\right]}. \tag{4.96}$$

Proof Given that $\mathbf{G_x}$ is a symmetric positive definite matrix, we can set the problem of bounding the maximum eigenvalue of $\mathbf{G_x}$ by 1 as [1]:

$$\max_{\mathbf{a}\in\mathbb{R}^L:\|\mathbf{a}\|=1} \mathbf{a}^T \mathbf{G_x}\mathbf{a} < 1. \tag{4.97}$$

Using (4.94) we get:

$$\|\mathbf{a}\|^2 - \mu\mathbf{a}^T (2\mathbf{R_x} - \mu\mathbf{S_x})\,\mathbf{a} < 1, \quad \forall\,\mathbf{a} \in \mathbb{R}^L, \quad \|\mathbf{a}\| = 1, \tag{4.98}$$

or equivalently

$$-\mu\mathbf{a}^T (2\mathbf{R_x} - \mu\mathbf{S_x})\,\mathbf{a} < 0, \quad \forall\,\mathbf{a} \in \mathbb{R}^L, \quad \|\mathbf{a}\| = 1. \tag{4.99}$$

---

[13] We will use the usual partial ordering defined for symmetric positive definite matrices [35].

[14] It is in this place where we only keep the sufficiency and lose the necessity.

From this equation we already see that, given that $\mu > 0$,

$$\mathbf{a}^T (2\mathbf{R_x} - \mu\mathbf{S_x}) \mathbf{a} > 0, \quad \forall \, \mathbf{a} \in \mathbb{R}^L, \quad \|\mathbf{a}\| = 1, \tag{4.100}$$

which means that the matrix $2\mathbf{R_x} - \mu\mathbf{S_x}$ has to be positive definite. This is equivalent to ask for $I - \frac{\mu}{2}\mathbf{R_x}^{-1/2}\mathbf{S_x}\mathbf{R_x}^{-1/2}$ to be positive definite. If this condition has to be satisfied, then the eigenvalues of $\frac{\mu}{2}\mathbf{R_x}^{-1/2}\mathbf{S_x}\mathbf{R_x}^{-1/2}$ must fulfill the following condition:

$$\frac{\mu}{2}\mathrm{eig}_i \left[\mathbf{R_x}^{-1/2}\mathbf{S_x}\mathbf{R_x}^{-1/2}\right] < 1, \quad i = 1, \ldots, L. \tag{4.101}$$

From this we can obtain the condition (4.96) using the fact that $\mathbf{R_x}^{-1}\mathbf{S_x}$ is *similar* [1] to $\mathbf{R_x}^{-1/2}\mathbf{S_x}\mathbf{R_x}^{-1/2}$, which implies that they have the same eigenvalues. □

To recapitulate, we can see that a sufficient condition for the MSD stability of the LMS algorithm is given by (4.96). In order to solve this problem we need to know the matrix $\mathbf{S_x}$. Although in general this quantity cannot be known in closed form, we can always resort to a numerical integration routine if we know the pdf of $\mathbf{x}(n)$. However, from (4.96) we can get some intuition about the maximum value of μ. Consider the following quantity:

$$\mathbf{T_x} = E \left[\frac{1}{L}\|\mathbf{x}(n)\|^2 \mathbf{x}(n)\mathbf{x}^T(n)\right] = \frac{\mathbf{S_x}}{L}. \tag{4.102}$$

Supposing that $\mathbf{x}(n)$ is constructed from a scalar stationary process $x(n)$, under very mild conditions [38], it can be proved that the variance of $\frac{1}{L}\|\mathbf{x}(n)\|^2$ is such that it decreases with K/L, for some finite value $K > 0$. That is, as L grows the value of $\frac{1}{L}\|\mathbf{x}(n)\|^2$ is more similar to its mean value σ_x^2, and its dispersion around that value decreases. This allows us to simplify (4.102) as:

$$\mathbf{T_x} \approx \sigma_x^2 E\left[\mathbf{x}(n)\mathbf{x}^T(n)\right], \tag{4.103}$$

or what is the same,

$$\mathbf{S_x} \approx L\sigma_x^2 \mathbf{R_x} = \mathrm{tr}\,[\mathbf{R_x}]\,\mathbf{R_x}. \tag{4.104}$$

Then, we can see that if L is long enough:

$$\mathrm{eig}_{\max}\left[\mathbf{R_x}^{-1}\mathbf{S_x}\right] \approx L\sigma_x^2 = \mathrm{tr}\,[\mathbf{R_x}]. \tag{4.105}$$

Therefore, for sufficiently long filters we obtain that the limit for MSD stability is approximately:

$$\mu \approx \frac{2}{L\sigma_x^2} = \frac{2}{\mathrm{tr}\,(\mathbf{R_x})}. \tag{4.106}$$

The important thing here is that the stability limit decreases with the increasing length of the adaptive filter. That is, longer filters will have a more restricted interval of valid μ values. Although this was derived from condition (4.96), which is valid under the independence assumption, the result in (4.106) (i.e., that the maximum μ for which the filter remains stable is inversely proportional to the filter length) is observed to be true in general, even when that assumption is not satisfied. The reader should be cautioned about the last discussion. Although our conclusion was derived in an intuitive way, it can be justified in rigorous mathematical terms. However, those mathematical developments, although important on their own as Theorem 4.1, are not directly linked with our interest in adaptive filters, and for this reason we chose to present the above important result in the way we did.

In the case where the input vectors pdf is Gaussian with covariance matrix $\mathbf{R_x}$, we can obtain an explicit and exact bound for μ. We can use the result known as Gaussian factoring theorem[15] for the obtention of $\mathbf{S_x}$, leading to:

$$\mathbf{S_x} = 2\mathbf{R_x^2} + \mathbf{R_x}\text{tr}\,[\mathbf{R_x}] .\qquad(4.107)$$

Then, using (4.96) we can show that:

$$0 < \mu < \frac{2}{2\,\text{eig}_{\max}\,[\mathbf{R_x}] + \text{tr}\,[\mathbf{R_x}]}.\qquad(4.108)$$

This a sufficient condition for the MSD stability of the LMS with Gaussian and independent input regressors. However, it is close enough to the necessary and sufficient condition that can be obtained from a more elaborated model of the transient behavior of the LMS algorithm under Gaussian signaling [19, 40]. In fact, for white Gaussian signals condition (4.108) is also necessary.

A more restrictive but useful bound for the stability of the LMS can be obtained from the fact that $\text{eig}_{\max}\,[\mathbf{R_x}] \le \text{tr}\,[\mathbf{R_x}]$, and therefore

$$0 < \mu < \frac{2}{3\text{tr}\,[\mathbf{R}_x]}.\qquad(4.109)$$

This bound was obtained with a different approach in [31]. Obviously the case $\mu = 0$, although it gives $\text{eig}_{\max}\,[\mathbf{G_x}] = 1$, is a stable point. However, it is useless because it means the algorithm will not move and stay forever at the initial condition. The reader can also see that equation (4.109) has the same asymptotic behavior as (4.106). In the general non-Gaussian case, the sufficient condition on μ for MSD stability has to be derived from (4.96).

---

[15] For Gaussian random variables we have the following result [39]:

$$E[x_1x_2x_3x_4] = E[x_1x_2]E[x_3x_4] + E[x_1x_3]E[x_2x_4] + E[x_1x_4]E[x_2x_3].$$

- **NLMS algorithm**: For the MSD stability of the NLMS algorithm we need to consider the matrix $\mathbf{G_x}$, that in this case can be written as:

$$\mathbf{G_x} = \mathbf{I}_L - 2\mu\mathbf{K_x} + \mu^2\mathbf{H_x}, \tag{4.110}$$

where

$$\mathbf{H_x} = E\left[\frac{\|\mathbf{x}(n)\|^2}{\left(\|\mathbf{x}(n)\|^2 + \delta\right)^2}\mathbf{x}(n)\mathbf{x}^T(n)\right]. \tag{4.111}$$

By applying Lemma 4.3, the resulting stable range of μ is

$$0 < \mu < \frac{2}{\text{eig}_{\max}\left[\mathbf{K_x}^{-1}\mathbf{H_x}\right]}. \tag{4.112}$$

For the case $\delta = 0$, it is easy to see that $\mathbf{H_x} = \mathbf{K_x}$, leading to

$$0 < \mu < 2. \tag{4.113}$$

The condition on the step size of the NLMS algorithm for mean square stability is independent of any statistical characteristic of the input vector and is not restricted to the Gaussian case only. This is the same sufficient condition for the MSD stability of the NLMS algorithm obtained in [40–42], and [43], where specific distributions for the input vector were used. Here we do not use any statistical information about the input vectors; just the independence assumption which is used in those works as well. In the case $\delta \neq 0$, the stability bound (4.113) is still sufficient. Notice, that in contrast to the LMS algorithm, the stability result for the NLMS algorithm does not depend on the filter length L.

- **Sign Data algorithm**: Matrix $\mathbf{G_x}$ can be written for the SDA as:

$$\mathbf{G_x} = \mathbf{I}_L - 2\mu\mathbf{L_x} + \mu^2 L\mathbf{R_x}, \tag{4.114}$$

where we used the fact that $E\left[\|\text{sign}\left[\mathbf{x}(n)\right]\|^2\mathbf{x}(n)\mathbf{x}^T(n)\right] = L\mathbf{R_x}$. Application of Lemma 4.3 leads us to the following bound for the MSD stability:

$$0 < \mu < \frac{2}{L\,\text{eig}_{\max}\left[\mathbf{L_x}^{-1}\mathbf{R_x}\right]}. \tag{4.115}$$

In order to obtain a more explicit result we need to find $\mathbf{L_x}$. A closed form expression of this quantity is not possible in general, but we can always use computer simulations to have and estimate. However, using the expression for $\mathbf{L_x}$ obtained in (4.78), for the important Gaussian case (4.115) can be put as

$$0 < \mu < \sqrt{\frac{8}{\pi \sigma_x^2} \frac{1}{L}}, \qquad (4.116)$$

which is the same result obtained in [14] and [40]. From the results in [14], where this stability bound is calculated based on a full transient analysis, we know that this bound is also a necessary condition for convergence. These results, and the ones corresponding to the mean stability of the SDA, are valid if $\mathbf{L_x}$ is a positive definite matrix. If this is not the case (for example if some of the eigenvalues of $\mathbf{L_x}$ have negative real parts), it would be possible that no choice of μ will guarantee the stability of the algorithm (in mean and MSD sense). This will be the case for some class of input signals as it is shown in [44]. Moreover, certain signals might lead to stability of the LMS algorithm but not of the SDA [13].

4.5.3 Steady State Behavior for a Large Family of Algorithms

Once we have conditions for the mean and MSD stability of an adaptive filter algorithm we could get some insight into its asymptotic or steady state behavior. The approach we take in this section will allow us to obtain the asymptotic value of:

$$E\left[e^2(n)\right] = E\left[\left(\tilde{\mathbf{w}}^T(n-1)\mathbf{x}(n) + v(n)\right)^2\right]$$

$$= E\left[\left(\tilde{\mathbf{w}}^T(n-1)\mathbf{x}(n)\right)^2\right] + \sigma_v^2$$

$$= \xi(n) + \sigma_v^2, \qquad (4.117)$$

where we used the assumptions about the noise sequence, the fact that the a priori error can be written as $\eta(n) = \tilde{\mathbf{w}}^T(n-1)\mathbf{x}(n)$, and that the EMSE is $\xi(n) = E\left[\eta^2(n)\right]$. Note that if the value of $\lim_{n\to\infty} \xi(n)$ is different from zero, the final MSE will be greater than σ_v^2, which corresponds to the optimal Wiener error (as discussed in Sect. 2.4). This can be interpreted as follows: once the adaptive filter has reached its steady state, it will keep on moving with random fluctuations in a vicinity of the optimal Wiener solution. The amplitude of these fluctuations is measured by the EMSE or the final MSD. Note that if we consider the independence assumption between the input vectors, we have:

$$\xi(n) = \mathrm{tr}\left[\mathbf{R_x}\mathbf{K}(n-1)\right], \quad \text{with} \quad \mathbf{K}(n-1) = E\left[\tilde{\mathbf{w}}(n-1)\tilde{\mathbf{w}}^T(n-1)\right], \quad (4.118)$$

Here, we used the result (4.50), given the fact that $\tilde{\mathbf{w}}(n-1)$ is a function of the past values of the input vectors and the noise sequence. This means, that we can bound the EMSE as[16]:

$$\text{eig}_{\min}\,[\mathbf{R_x}]\,E\left[\|\tilde{\mathbf{w}}(n-1)\|^2\right] \le \xi(n) \le \text{eig}_{\max}\,[\mathbf{R_x}]\,E\left[\|\tilde{\mathbf{w}}(n-1)\|^2\right], \quad (4.119)$$

where we used the fact that $\text{tr}\,[\mathbf{K}(n-1)] = E\left[\|\tilde{\mathbf{w}}(n-1)\|^2\right]$. Equation (4.119) is also valid when $n \to \infty$, provided that the adaptive filter is stable in the MSD sense. This means that the steady state MSD will have almost the same behavior with respect to the step size as the final EMSE (at least we can bound it uniformly from below and above with respect to the EMSE behavior as a function of the step size). If the input is such that $\mathbf{R_x} = \sigma_x^2 \mathbf{I}_L$,

$$\xi(n) = \sigma_x^2 E\left[\|\tilde{\mathbf{w}}(n-1)\|^2\right], \quad (4.120)$$

which means that for white input regressors we can know exactly the behavior of the final MSD from the asymptotic behavior of $\xi(n)$.

The interesting thing however, is that for the final EMSE we do not need to resort to the full independence assumption between the input regressors that was used for the MSD stability analysis. In order to provide a unified treatment to the steady state behavior and make the results valid for a large class of algorithms, we will continue with the general model assumed in (4.55), which is reproduced here in the following equivalent form:

$$\tilde{\mathbf{w}}(n) = \alpha\tilde{\mathbf{w}}(n-1) - \boldsymbol{\mu}\mathbf{f}\,(\mathbf{x}(n))\,e(n) + (1-\alpha)\mathbf{w}_T. \quad (4.121)$$

Squaring both sides we obtain:

$$\begin{aligned}
\|\tilde{\mathbf{w}}(n)\|^2 = {}& \alpha^2\|\tilde{\mathbf{w}}(n-1)\|^2 - 2\alpha\tilde{\mathbf{w}}^T(n-1)\boldsymbol{\mu}\mathbf{f}\,(\mathbf{x}(n))\,e(n) \\
& + 2\alpha(1-\alpha)\mathbf{w}_T^T\tilde{\mathbf{w}}(n-1) \\
& + \mathbf{f}^T\,(\mathbf{x}(n))\,\boldsymbol{\mu}^2\mathbf{f}\,(\mathbf{x}(n))\,e^2(n) - 2(1-\alpha)\mathbf{f}^T\,(\mathbf{x}(n))\,\boldsymbol{\mu}\mathbf{w}_T e(n) \\
& + (1-\alpha)^2\|\mathbf{w}_T\|^2. \quad (4.122)
\end{aligned}$$

If we define $\eta^{\mu f}(n) = \tilde{\mathbf{w}}^T(n-1)\boldsymbol{\mu}\mathbf{f}\,(\mathbf{x}(n))$ and apply the expectation operator on both sides we obtain:

$$\begin{aligned}
E\left[\|\tilde{\mathbf{w}}(n)\|^2\right] = {}& \alpha^2 E\left[\|\tilde{\mathbf{w}}(n-1)\|^2\right] - 2\alpha E\left[\eta^{\mu f}(n)e(n)\right] \\
& + 2\alpha(1-\alpha)\mathbf{w}_T^T E\left[\tilde{\mathbf{w}}(n-1)\right] + E\left[\|\mathbf{f}^T\,(\mathbf{x}(n))\,\|_{\boldsymbol{\mu}^2}^2 e^2(n)\right] \\
& - 2(1-\alpha)E\left[\mathbf{f}^T\,(\mathbf{x}(n))\,\boldsymbol{\mu}\mathbf{w}_T e(n)\right] + (1-\alpha)^2\|\mathbf{w}_T\|^2, \quad (4.123)
\end{aligned}$$

---

[16] For this, we need $\mathbf{R_x}$ to be strictly positive definite, which was assumed in Sect. 4.5.1.

where $\|\mathbf{f}(\mathbf{x}(n))\|_{\mu^2}^2 = \mathbf{f}^T(\mathbf{x}(n))\,\mu^2\mathbf{f}(\mathbf{x}(n))$. It is clear that (4.123) is an exact equality and that no particular assumptions have been taken. If μ is chosen in order to provide mean and MSD stability, we can assume that the limit $\lim_{n\to\infty} E\left[\|\tilde{\mathbf{w}}(n)\|^2\right]$ exists. In an abuse of notation, this quantity will be denoted as $E\left[\|\tilde{\mathbf{w}}(\infty)\|^2\right]$. Taking limits on both sides of (4.123) and rearranging terms leads to

$$(1 - \alpha^2)E\left[\|\tilde{\mathbf{w}}(\infty)\|^2\right] = \lim_{n\to\infty}\left\{-2\alpha E\left[\eta^{\mu f}(n)e(n)\right] + 2\alpha(1 - \alpha)\mathbf{w}_T^T E\left[\tilde{\mathbf{w}}(n - 1)\right]\right.$$

$$\left. + E\left[\|\mathbf{f}(\mathbf{x}(n))\|_{\mu^2}^2 e^2(n)\right] - 2(1 - \alpha)E\left[\mathbf{f}^T(\mathbf{x}(n))\,\mu\mathbf{w}_T e(n)\right] + (1 - \alpha)^2\|\mathbf{w}_T\|^2\right\}.$$

$$(4.124)$$

Now we analyze some of the terms in (4.124). The term $E\left[\mathbf{f}^T(\mathbf{x}(n))\,\mu\mathbf{w}_T e(n)\right]$ can be put as:

$$E\left[\mathbf{f}^T(\mathbf{x}(n))\,\mu\mathbf{w}_T e(n)\right] = E\left\{\mathbf{f}^T(\mathbf{x}(n))\,\mu\mathbf{w}_T\,[\eta(n) + v(n)]\right\}$$

$$= E\left[\mathbf{f}^T(\mathbf{x}(n))\,\mu\mathbf{w}_T\,\eta(n)\right] + E\left[\mathbf{f}^T(\mathbf{x}(n))\,\mu\mathbf{w}_T v(n)\right].$$

$$(4.125)$$

Because of the independence between the input and the noise and the fact that the noise has zero mean, $E\left[\mathbf{f}^T(\mathbf{x}(n))\,\mu\mathbf{w}_T v(n)\right] = 0$. The term $E\left[\mathbf{f}^T(\mathbf{x}(n))\,\mu\mathbf{w}_T\,\eta(n)\right]$ can also be written as:

$$E\left[\mathbf{f}^T(\mathbf{x}(n))\,\mu\mathbf{w}_T\,\eta(n)\right] = E_{\mathbf{x}(n)}\left[\mathbf{f}^T(\mathbf{x}(n))\,\mu\mathbf{w}_T\,E_{\eta(n)|\mathbf{x}(n)}\,[\eta(n)|\mathbf{x}(n)]\right].\quad(4.126)$$

When $n \to \infty$, we have $E_{\eta(n)|\mathbf{x}(n)}\,[\eta(n)|\mathbf{x}(n)] \approx E\,[\eta(n)]$. The reason for this is that when n is large and the algorithm is reaching its steady state, successive values of the input vectors will not influence significantly $\tilde{\mathbf{w}}(n)$ and $\eta(n)$ (on average). This means that for large n, $\eta^2(n)$ is mean square unpredictable (MSU) with respect to $\mathbf{x}(n)$ and that in some sense there is no information in $\mathbf{x}(n)$ which is relevant to explain the fluctuations of $\eta(n)$. A similar conclusion can be obtained if we remember that $\eta(n) = \tilde{\mathbf{w}}^T(n - 1)\mathbf{x}(n)$ and that after a sufficient long time $\tilde{\mathbf{w}}(n - 1)$ will be small (because $\mathbf{w}(n - 1)$ will be "close" to $\mathbf{w}_T$) with respect to the values of $\mathbf{x}(n)$. This translates into the fact that on average, the value of $\eta(n)$ will have small variations independently of the variations on $\mathbf{x}(n)$. Then, it agrees with the intuitive reasoning given above that in the steady state, the input regressors will not have (on average) much influence on $\eta(n)$. This is less than asking for the independence between $\eta(n)$ and $\mathbf{x}(n)$ for large n [45], which would lead to same conclusion. Summarizing, we have for large n:

$$E\left[\mathbf{f}^T(\mathbf{x}(n))\,\mu\mathbf{w}_T e(n)\right] \approx \mathbf{w}_T^T\mu E\,[\mathbf{f}(\mathbf{x}(n))]\,E\,[\eta(n)].\quad(4.127)$$

In the same manner we have:

$$E\left[\|\mathbf{f}\left(\mathbf{x}(n)\right)\|_{\mu^2}^2 e^2(n)\right] = E\left[\|\mathbf{f}\left(\mathbf{x}(n)\right)\|_{\mu^2}^2 \left(\eta^2(n) + 2\eta(n)v(n) + v^2(n)\right)\right]$$
$$= E\left[\|\mathbf{f}\left(\mathbf{x}(n)\right)\|_{\mu^2}^2 \eta^2(n)\right] + \sigma_v^2 E\left[\|\mathbf{f}\left(\mathbf{x}(n)\right)\|_{\mu^2}^2\right],$$
$$\text{(4.128)}$$

where we used $E\left[\|\mathbf{f}\left(\mathbf{x}(n)\right)\|_{\mu^2}^2 \eta(n)v(n)\right] = 0$. With the reasoning used above we can write $E\left[\|\mathbf{f}\left(\mathbf{x}(n)\right)\|_{\mu^2}^2 \eta^2(n)\right] \approx E\left[\|\mathbf{f}\left(\mathbf{x}(n)\right)\|_{\mu^2}^2\right] E\left[\eta^2(n)\right]$, so (4.124) can be put as:

$$(1 - \alpha^2) E\left[\|\tilde{\mathbf{w}}(\infty)\|^2\right] \approx \lim_{n\to\infty} \left(-2\alpha E\left[\eta^{\mu f}(n)e(n)\right] + 2\alpha(1 - \alpha)\mathbf{w}_T^T E\left[\tilde{\mathbf{w}}(n - 1)\right]\right.$$
$$+ E\left[\|\mathbf{f}\left(\mathbf{x}(n)\right)\|_{\mu^2}^2\right]\xi(n) - 2(1 - \alpha)\mathbf{w}_T^T \mu E\left[\mathbf{f}\left(\mathbf{x}(n)\right)\right] E\left[\eta(n)\right]$$
$$\left. + \sigma_v^2 E\left[\|\mathbf{f}\left(\mathbf{x}(n)\right)\|_{\mu^2}^2\right] + (1 - \alpha)^2\|\mathbf{w}_T\|^2\right). \quad \text{(4.129)}$$

From this equation we can obtain steady state results for almost all the algorithms presented throughout this chapter.

- **LMS algorithm**: In this case $\eta^{\mu f}(n) = \mu\eta(n)$, so (4.129) takes the form[17]

$$0 = -2\mu\xi_\infty + \mu^2 E\left[\|\mathbf{x}(n)\|^2\right]\xi_\infty + \mu^2\sigma_v^2 E\left[\|\mathbf{x}(n)\|^2\right], \quad \text{(4.130)}$$

where we denoted $\xi_\infty = \lim_{n\to\infty}\xi(n)$. Rearranging terms in (4.130) we obtain:

$$\xi_\infty = \frac{\mu\,\text{tr}\,[\mathbf{R_x}]\,\sigma_v^2}{2 - \mu\,\text{tr}\,[\mathbf{R_x}]}, \quad \text{(4.131)}$$

where we have used that $E\left[\|\mathbf{x}(n)\|^2\right] = \text{tr}\,[\mathbf{R_x}]$. Notice that the final EMSE is increasing with μ and the variance of the noise. This means that in order to have a small EMSE the value of μ should be small. In fact, if μ is small the final EMSE can be approximated by:

$$\xi_\infty \approx \frac{\mu\,\text{tr}\,[\mathbf{R_x}]\,\sigma_v^2}{2}. \quad \text{(4.132)}$$

From (4.131) we can also see that there is a maximum value μ before ξ_∞ becomes infinite, and after which it would take negative values, which is meaningless. We see that this value of μ coincides with the approximate upperbound we derived in (4.106), which is satisfactory. Obviously, from (4.117) and (4.131) we also have:

$$J_\infty \triangleq \lim_{n\to\infty} E\left[|e(n)|^2\right] = \sigma_v^2 + \frac{\mu\,\text{tr}\,[\mathbf{R_x}]\,\sigma_v^2}{2 - \mu\,\text{tr}\,[\mathbf{R_x}]}. \quad \text{(4.133)}$$

---

[17] Although in (4.130) we should put $\approx$, in an abuse of notation we state it as an equality.

It is interesting to quantify how the value of J_∞ differs from the optimum value of the Wiener filtering J_{min}, which in this case is equal to σ_v^2. This leads us to the definition of the *misadjustment*:

$$\mathcal{M} \triangleq \frac{J_\infty - J_{min}}{J_{min}} = \frac{\xi_\infty}{J_{min}}. \tag{4.134}$$

For the LMS algorithm we obtain:

$$\mathcal{M} = \frac{\mu \operatorname{tr}[\mathbf{R_x}]}{2 - \mu \operatorname{tr}[\mathbf{R_x}]}, \tag{4.135}$$

which does not depend on the noise variance. It is not surprising to see that the misadjustment is increasing with μ.

Finally we can see that when the input is white, i.e., $\mathbf{R_x} = \sigma_x^2 \mathbf{I}_L$, from (4.120) it is easy to see that the final MSD is:

$$\lim_{n \to \infty} E\left[\|\tilde{\mathbf{w}}(n)\|^2\right] = \frac{\mu L \sigma_v^2}{2 - \mu L \sigma_x^2}, \tag{4.136}$$

which is an increasing function of μ and L.

- **NLMS algorithm**: We have $\eta^{\mu f}(n) = \frac{\mu}{\|\mathbf{x}(n)\|^2 + \delta} \eta(n)$. Equation (4.129) can be written as:

$$0 = -2\mu \lim_{n \to \infty} E\left[\frac{\eta^2(n)}{\|\mathbf{x}(n)\|^2 + \delta}\right] + \mu^2 \sigma_v^2 E\left[\frac{\|\mathbf{x}(n)\|^2}{(\|\mathbf{x}(n)\|^2 + \delta)^2}\right]$$

$$+ \mu^2 E\left[\frac{\|\mathbf{x}(n)\|^2}{(\|\mathbf{x}(n)\|^2 + \delta)^2}\right] \xi_\infty. \tag{4.137}$$

As reasoned before, we can make the following approximation:

$$E\left[\frac{\eta^2(n)}{\|\mathbf{x}(n)\|^2 + \delta}\right] \approx E\left[\frac{1}{\|\mathbf{x}(n)\|^2 + \delta}\right] \xi(n). \tag{4.138}$$

With (4.138) we can obtain ξ_∞ as:

$$\xi_\infty = \frac{\mu \sigma_v^2 E\left[\frac{\|\mathbf{x}(n)\|^2}{(\|\mathbf{x}(n)\|^2 + \delta)^2}\right]}{2E\left[\frac{1}{\|\mathbf{x}(n)\|^2 + \delta}\right] - \mu E\left[\frac{\|\mathbf{x}(n)\|^2}{(\|\mathbf{x}(n)\|^2 + \delta)^2}\right]}. \tag{4.139}$$

The quantities:

$$E\left[\frac{1}{\|\mathbf{x}(n)\|^2 + \delta}\right] \text{ and } E\left[\frac{\|\mathbf{x}(n)\|^2}{\left(\|\mathbf{x}(n)\|^2 + \delta\right)^2}\right], \tag{4.140}$$

can be evaluated with the input vector pdf. However, if $\delta = 0$, we have:

$$\xi_\infty = \frac{\mu \sigma_v^2}{2 - \mu}, \tag{4.141}$$

which, as in the stability condition seen before, does not depend on any characteristic of the input vectors pdf. The misadjustment for the NLMS algorithm in the case of $\delta = 0$ is:

$$\mathcal{M} = \frac{\mu}{2 - \mu}, \tag{4.142}$$

which depends only on μ. If we use the independence assumption and consider that $\mathbf{R_x} = \sigma_x^2 \mathbf{I}_L$, it is easy to show that:

$$\lim_{n \to \infty} E\left[\|\tilde{\mathbf{w}}(n)\|^2\right] = \frac{\mu}{2 - \mu} \frac{\sigma_v^2}{\sigma_x^2}. \tag{4.143}$$

If we cannot assume that $\mathbf{R_x} = \sigma_x^2 \mathbf{I}_L$, we can only bound the behavior of the asymptotic MSD as in (4.119). It is interesting to observe how the final EMSE and misadjustment reflect the stability bound $\mu < 2$ in their expressions, in a similar way as in the LMS algorithm.

As in the LMS case, even though these expressions have been derived under some particular assumptions, they have proved to be quite accurate when compared with results obtained from using the NLMS algorithm in real problems.

- **Sign Data algorithm**: We also have $\eta^{\mu f}(n) = \mu \tilde{\mathbf{w}}^T(n-1)\text{sign}\,[\mathbf{x}(n)]$. Equation (4.129) can be written as:

$$0 = -2\mu \lim_{n \to \infty} E\left[\tilde{\mathbf{w}}^T(n-1)\text{sign}\,[\mathbf{x}(n)]\,\mathbf{x}^T(n)\tilde{\mathbf{w}}(n-1)\right]$$
$$+ \mu^2 \sigma_v^2 E\left[\|\text{sign}[\mathbf{x}(n)]\|^2\right] + \mu^2 E\left[\|\text{sign}[\mathbf{x}(n)]\|^2\right]\xi_\infty. \tag{4.144}$$

Using that $\|\text{sign}\,[\mathbf{x}(n)]\|^2 = L$ we can write:

$$\xi_\infty = \frac{2\lim_{n \to \infty} E\left[\tilde{\mathbf{w}}^T(n-1)\text{sign}\,[\mathbf{x}(n)]\,\mathbf{x}^T(n)\tilde{\mathbf{w}}(n-1)\right] - \mu L \sigma_v^2}{\mu L}. \tag{4.145}$$

This is the general expression for the steady state behavior of the SDA. In order to obtain a more explicit expression we can assume that the input regressors are Gaussian, and the fact that when n is large $\tilde{\mathbf{w}}(n-1)$ is MSU with respect to $\mathbf{x}(n)$ (using the same reasoning presented before). With these assumptions,

$$E\left[\tilde{\mathbf{w}}^T(n-1)\text{sign}\left[\mathbf{x}(n)\right]\mathbf{x}^T(n)\tilde{\mathbf{w}}(n-1)\right] = \sqrt{\frac{2}{\pi\sigma_x^2}}E\left[\tilde{\mathbf{w}}^T(n-1)\mathbf{R_x}\tilde{\mathbf{w}}(n-1)\right]$$

$$= \sqrt{\frac{2}{\pi\sigma_x^2}}\xi(n). \qquad (4.146)$$

Combining the results from (4.145) and (4.146) we obtain:

$$\xi_\infty = \frac{\mu L\sigma_v^2}{\sqrt{\frac{8}{\pi\sigma_x^2}} - \mu L}. \qquad (4.147)$$

If we do not assume that $\mathbf{x}(n)$ is Gaussian, but restrict ourselves to the case where $\mathbf{R_x} = \sigma_x^2\mathbf{I}_L$ we obtain:

$$\xi_\infty = \frac{\mu L\sigma_v^2}{2\frac{E[|x|]}{\sigma_x^2} - \mu L}, \qquad (4.148)$$

where $E[|x|]$ is the expectation of the absolute value of any entry of vector $\mathbf{x}(n)$.

4.5.4 Transient and Tracking Behavior for Adaptive Filters

It was mentioned above that the analysis of the transient behavior of an adaptive filter is an important issue to be considered. It is through the transient behavior of an adaptive filter (mean transient or mean square transient behavior) that a fundamental tradeoff is unveiled: speed of convergence versus steady state error, and more importantly its relation to the step size μ. Then, depending on the final application, some important constraints on the steady state error or the convergence speed of an adaptive algorithm can be placed. Unfortunately, the obtention of closed form results about this fundamental tradeoff is very difficult from the mathematical point of view, without assuming stronger assumptions about the input process [19]. This is specially true when the interest is in the mean square transient behavior. There are however, interesting approaches to obtain results about the transient behavior of general adaptive filters, which in general lead to expressions that can be evaluated numerically [40, 45].

As in the case of the SD, the mean and mean square transient behavior of an adaptive algorithm is determined by different convergence modes, all of them related to the eigenvalues of the input correlation matrix $\mathbf{R_x}$. Depending on the values of μ and the eigenvalues of $\mathbf{R_x}$ some modes can converge faster than others. In general, as the dispersion of the eigenvalues is greater, the convergence speed of the algorithm will be affected. This dispersion on the eigenvalues of $\mathbf{R_x}$, captured by its condition number $\chi(\mathbf{R_x})$, translates into the *color* of the input signal. As the input signal is more similar to white noise, i.e., $\mathbf{R_x} = \sigma_x^2\mathbf{I}_L$ and $\chi(\mathbf{R_x}) = 1$, the dispersion is minimal and the convergence is the fastest. When the input signal is more *colored*, the dispersion between the eigenvalues is greater, i.e., $\mathbf{R_x} \neq \sigma_x^2 \mathbf{I}_L$ and $\chi\,(\mathbf{R_x}) \gg 1$, and the convergence becomes slower. This characteristic is the same we saw in the SD algorithm. As many real world signals are highly colored, e.g. speech signals, the slow convergence might preclude some algorithms from being used in some applications. In order to improve the convergence behavior of an adaptive algorithm working with this kind of signals, some *pre-whitening* of the input signal can be done before it enters into the adaptive filter [46, 47] (using for example linear prediction as in Sect. 2.5.1). In fact, other adaptive algorithms are better suited for colored signals as the Affine Projection algorithm (APA) and the Recursive Least Squares (RLS) algorithm. Both algorithms will be presented later.

Also, general stochastic gradient adaptive algorithms present the behavior mentioned in Sect. 4.2.2. That is, as the value of μ decreases, the steady state error gets smaller and the time required to reach that steady state gets larger, i.e., the speed of convergence is reduced. In the same manner, as μ increases, the steady state error and the speed of convergence increase as well. The behavior mentioned above, happens $\forall \mu < \mu^*$, where the exact value of μ^* depends on the particular algorithm considered. When μ is increased above μ^* the speed of convergence could be reduced and the steady state error is increased. From this behavior we conclude that in practice, the best of an adaptive algorithm can be obtained restricting the choice of the step size to $(0, \mu^*]$. The obtention of μ^* is difficult, not only for the mathematics and assumptions needed to obtain a closed form expression, but also for the criterion used to quantify the convergence speed of an adaptive algorithm. A reasonable criterion would be to obtain the value of μ such that for the slower mode its convergence speed is maximized, as we did for the SD algorithm in Sect. 3.2.1. When the same is done for the mean square behavior of the LMS, assuming that $\mathbf{R_x} = \sigma_x^2 \mathbf{I}_L$ and the input is Gaussian, it can be shown that

$$\mu^* = \frac{1}{2\sigma_x^2 + L\sigma_x^2} \tag{4.149}$$

which is the midpoint of the corresponding stability interval computed using (4.108). Increasing μ above that value, will decrease the convergence speed of the slowest mode and increase the steady state error. Decreasing μ below μ^* will also decrease the convergence speed of the slowest mode but will decrease the steady state error. This confirms, for this case, the fact that $\mu \in (0, \mu^*]$ is the most useful range of μ as mentioned above. With extra effort, it can be proved that under the same conditions and with sufficiently large L, the value of μ^* for the NLMS algorithm is close to one, confirming also the intuitive and heuristic discussion in Sect. 4.2.2. In this way, for the NLMS algorithm, the most useful range of μ is $(0, 1]$.

The tracking behavior of an adaptive filter is another important issue to be considered. As an adaptive algorithm works with the instantaneous data, it can potentially track the statistical variations on it. This is one of the most useful properties of an adaptive filter. In our analyses we considered that the signals are stationary and the system $\mathbf{w}_T$ does not vary with time. In practice, this is generally not true. For example, in an acoustic echo cancelation application, the input signals are speech signals, which are nonstationary signals, and the room acoustic impulsive response (the system to be identified by the adaptive filter) could be a time variant linear system [47]. This leads to the issue of how well an adaptive algorithm is able to track and adapt to these variations. It can be shown [19, 45] that, as in the case of the transient behavior, there is also a tradeoff between the steady state behavior and the tracking behavior. As μ increases, the algorithm is able to react faster to changes in the signal statistics or to changes in the system $\mathbf{w}_T$ to be identified, but its steady state performance worsens. Several interesting approaches [48] exist to calculate expressions where this tradeoff can be quantified without needing to resort to a full transient analysis. In fact, the approach taken in Sect. 4.5.3 can be easily modified to include the tracking behavior of a large family of stochastic gradient adaptive filters. For more in depth discussions of this important topic, the interested reader can see [19, 45].

4.6 Affine Projection Algorithm

From (4.23) it has been seen that when $\mu = 1$ and $\delta = 0$, the NLMS projects at each time step the weight error vector onto the orthogonal space spanned by the regressor $\mathbf{x}(n)$. The projector has $L - 1$ eigenvalues equal to one and the remaining one is equal to zero. Now, what would happen if the projector has K eigenvalues equal to zero and $L - K$ equal to one? In this case, extending the discussion from Sect. 4.2.2, it can be expected an increase on the average speed of convergence. This is the main idea behind the *Affine Projection algorithm* (APA), originally introduced in [49]. The cost behind this is an increase in computational complexity and possibly steady state error. However, when colored input signals are used, the convergence speed of LMS and NLMS could be so deteriorated [45] that the APA appears as a very appealing alternative.

4.6.1 APA as the Solution to a Projection Problem

First, we introduce the $L \times K$ data matrix $\mathbf{X}(n) = [\mathbf{x}(n) \ \mathbf{x}(n-1) \ \cdots \ \mathbf{x}(n-K+1)]$, the $K \times 1$ desired output vector $\mathbf{d}(n) = [d(n) \ d(n-1) \ \cdots \ d(n-K+1)]^T$, the a priori[18] output estimation error vector $\mathbf{e}(n) = \mathbf{d}(n) - \mathbf{X}^T(n)\mathbf{w}(n-1)$, and the a posteriori output estimation error vector $\mathbf{e}_p(n) = \mathbf{d}(n) - \mathbf{X}^T(n)\mathbf{w}(n)$.

Following the same ideas used for the NLMS, the regularized APA comes as the solution to the problem of finding the estimate $\mathbf{w}(n)$ that solves:

---

[18] It should be emphasized that $\|\mathbf{e}(n)\|^2$ is not the same as the sum of the squares of the last K output estimation errors, $\{e(i)\}_{i=n-K+1}^{n}$. Each component of the vector $\mathbf{e}(n)$ is computed using the same filter estimate $\mathbf{w}(n-1)$.

$$\min_{\mathbf{w}(n)} \|\mathbf{w}(n) - \mathbf{w}(n-1)\|^2 \text{ subject to } \mathbf{e}_p(n) = \left[\mathbf{I}_K - \mu \mathbf{X}^T(n)\mathbf{X}(n)\mathbf{S}_\delta(n) \right] \mathbf{e}(n),$$

$$(4.150)$$

where $\mathbf{S}_\delta(n) = \left[\delta \mathbf{I}_K + \mathbf{X}^T(n)\mathbf{X}(n) \right]^{-1}$. The solution to this optimization problem is:

$$\mathbf{w}(n) = \mathbf{w}(n-1) + \mu \mathbf{X}(n) \left[\delta \mathbf{I}_K + \mathbf{X}^T(n)\mathbf{X}(n) \right]^{-1} \mathbf{e}(n), \mathbf{w}(-1). \qquad (4.151)$$

Therefore, the APA computes the new estimate at each time step by doing the orthogonal projection of the old estimate onto the plane generated by the constraint in (4.150). When $\mu = 1$ and $\delta = 0$, the projection is done onto the space $\mathbf{e}_p(n) = 0$. Then, the new estimate is the best one (in terms of squared error) to explain the observation $\mathbf{d}(n)$ based on the input data $\mathbf{X}(n)$. When the observations are noisy, the algorithm will be compensating the effect of the noise, introducing an error in the estimation of the unknown system. To diminish this, $\mu < 1$ is usually used.

The LMS and NLMS algorithms use only the most recent regressor to update the weight vector estimate, whereas APA uses the K most recent regressors. Actually, when $K = 1$, (4.151) is equivalent to the regularized NLMS. For this reason, affine projection algorithms are sometimes called data reusing algorithms since they reuse past regressors and reference data (for example, the binormalized data reusing LMS algorithm [50] corresponds to the special case of an APA with $K = 2$). The integer K is referred to as the order of the APA algorithm.

Assuming $\mu = 1$ and $\delta = 0$ in (4.151) and replacing the output estimation error vector, the APA recursion can be put as:

$$\mathbf{w}(n) = [\mathbf{I}_L - \mathbf{P}(n)] \mathbf{w}(n-1) + \mathbf{X}(n) \left[\mathbf{X}^T(n)\mathbf{X}(n) \right]^{-1} \mathbf{d}(n), \qquad (4.152)$$

where the projector is defined as $\mathbf{P}(n) = \mathbf{X}(n) \left[\mathbf{X}^T(n)\mathbf{X}(n) \right]^{-1} \mathbf{X}^T(n)$.[19] Therefore, the APA first projects the estimate at time $n - 1$ onto the orthogonal space spanned by the K most recent regressors. Then, the second term in (4.152) is added, which turns the projection into an affine one.

When $\mu = 1$ and $\delta = 0$ the APA must satisfy the constraint $\mathbf{e}_p(n) = 0$, i.e., $\mathbf{d}(n) = \mathbf{X}^T(n)\mathbf{w}(n)$. When $K = 1$, the infinitely many vectors satisfying this constraint lie on an affine subspace denoted by $\mathscr{A}_n$. This space is affine because the null vector $\mathbf{0}$ may not belong to the space. The NLMS algorithm will therefore be updated to the point in this subspace with the minimum distance from the previous estimate $\mathbf{w}(n-1)$. When $K > 1$, the constraint can be splitted into K different constraints:

$$d(n-k) = \mathbf{x}^T(n-k)\mathbf{w}(n), \quad k = 0, 1, \dots, K-1. \qquad (4.153)$$

---

[19] In Chap. 5 we will provide a deeper discussion about the properties of orthogonal projection operators.

For each k, the infinitely many vectors satisfying the constraint lie on an affine subspace denoted by $\mathscr{A}_{n-k}$. This space will be orthogonal to $\mathbf{x}(n-k)$. The APA of order K will then find the update by projecting onto the intersection of the K affine subspaces. It should be noticed that in general the projection order satisfies $L \gg K$.

Finally, it is clear that the update will be found in $\mathscr{R}[\mathbf{X}(n)]$, the range or column space of $\mathbf{X}(n)$. This space can also be generated by choosing for $1 \leq k \leq K$ the component from each vector $\mathbf{x}(n-k)$ that is orthogonal to $\mathbf{x}(n), \mathbf{x}(n-1), \ldots, \mathbf{x}(n-k+1)$. To do this, Gram-Schmidt orthogonalization can be used, which is the approach chosen for the NLMS-OCF algorithm [51].

4.6.2 Computational Complexity of APA

Assuming that the cost of inverting an $K \times K$ matrix is $O\left(K^3\right)$, the APA is $O\left(K^2 L\right)$.[20] Although this is larger than the cost of the LMS or NLMS, fast versions of the APA have been proposed. By looking at the temporal structure of the input data, the redundancies between successive samples can be exploited. One example is the Fast Affine Projection (FAP) algorithm [52], which performs $2L + O(K)$ operations per iteration, so with $K \ll L$, the computational cost will be close to $O(L)$. Moreover, since FAP algorithms require the solution of a linear system of the form $\boldsymbol{\epsilon}(n) = \mathbf{R}^{-1}(n)\mathbf{e}(n)$, different implementations will arise depending on how this system is solved [53]. Block processing is also used to save computational cost. One example is the Block Exact Fast Affine Projection (BEFAP) [54]. The main idea is to compute all the required quantities to perform the filter update only every K iterations. This kind of solutions are generally implemented in the frequency domain using fast convolution schemes.

4.6.3 Decorrelating Property of APA

In [42], the study on the convergence of LMS and NLMS algorithms showed that the latter presents a faster speed of convergence but with an increase in the misadjustment. Let us assume $\mu = 1$ and highly correlated input regressors. Then, at two consecutive time steps, the resulting affine subspaces associated with the constraints (4.153) will be very similar. Therefore, it can be expected that the magnitude of the NLMS update will be small, resulting in a slow convergence.

If $\mu = 1$ and $\delta = 0$, the APA recursion results in:

$$\mathbf{w}(n) = \mathbf{w}(n-1) + \mathbf{X}(n)\mathbf{S}(n)\mathbf{e}(n), \quad \mathbf{w}(-1), \tag{4.154}$$

---

[20] Usual matrix inverting algorithms as Gaussian elimination require basically K^3 multiplications [1]. The cost of APA is dominated by the $K^2 L$ multiplications and $K^2(L-1)$ additions required to compute $\mathbf{X}^T(n)\mathbf{X}(n)$, since with $L \gg K$ the cost of matrix inversion becomes less important.

where $\mathbf{S}(n) = \left[\mathbf{X}^T(n)\mathbf{X}(n)\right]^{-1}$. Under these conditions, it can be seen that the last $K - 1$ components of the vector $\mathbf{e}(n)$ will be zero (as they are the first $K - 1$ components of $\mathbf{e}_p(n - 1)$). Therefore, only the first column of $\mathbf{S}(n)$ will be relevant for the update.

Now, let the $L \times (K - 1)$ matrix $\tilde{\mathbf{X}}(n)$ include the input data vectors from time $n - 1$ to $n - K + 1$, i.e.,

$$\mathbf{X}(n) = \left[\mathbf{x}(n) \; \tilde{\mathbf{X}}(n)\right]. \tag{4.155}$$

Using (4.155), the matrix $\mathbf{S}(n)$ can be written as:

$$\mathbf{S}(n) = \begin{bmatrix} s(n) & \mathbf{s}^T(n) \\ \mathbf{s}(n) & \tilde{\mathbf{S}}(n) \end{bmatrix} = \begin{bmatrix} \mathbf{x}^T(n)\mathbf{x}(n) & \mathbf{x}^T(n)\tilde{\mathbf{X}}(n) \\ \tilde{\mathbf{X}}^T(n)\mathbf{x}(n) & \tilde{\mathbf{X}}^T(n)\tilde{\mathbf{X}}(n) \end{bmatrix}^{-1}, \tag{4.156}$$

and the APA update takes the form:

$$\mathbf{X}(n)\mathbf{S}(n)\mathbf{e}(n) = \left[\mathbf{x}(n)s(n) + \tilde{\mathbf{X}}(n)\mathbf{s}(n)\right]e(n). \tag{4.157}$$

In (4.156), the inverse on the right can be solved in terms of the original blocks, leading to:

$$s(n) = \frac{1 - \mathbf{x}^T(n)\tilde{\mathbf{X}}(n)\mathbf{s}(n)}{\mathbf{x}^T(n)\mathbf{x}(n)},$$

$$\mathbf{s}(n) = \frac{-\left[\tilde{\mathbf{X}}^T(n)\tilde{\mathbf{X}}(n)\right]^{-1}\tilde{\mathbf{X}}^T(n)\mathbf{x}(n)}{\mathbf{x}^T(n)\left\{\mathbf{I}_L - \tilde{\mathbf{X}}(n)\left[\tilde{\mathbf{X}}^T(n)\tilde{\mathbf{X}}(n)\right]^{-1}\tilde{\mathbf{X}}^T(n)\right\}\mathbf{x}(n)} \tag{4.158}$$

Replacing in (4.157), the APA update can be put as:

$$\left[\mathbf{x}(n)s(n) + \tilde{\mathbf{X}}(n)\mathbf{s}(n)\right]e(n) = \frac{\left\{\mathbf{I}_L - \tilde{\mathbf{X}}(n)\left[\tilde{\mathbf{X}}^T(n)\tilde{\mathbf{X}}(n)\right]^{-1}\tilde{\mathbf{X}}^T(n)\right\}\mathbf{x}(n)e(n)}{\mathbf{x}^T(n)\left\{\mathbf{I}_L - \tilde{\mathbf{X}}(n)\left[\tilde{\mathbf{X}}^T(n)\tilde{\mathbf{X}}(n)\right]^{-1}\tilde{\mathbf{X}}^T(n)\right\}\mathbf{x}(n)}$$

$$= \frac{\mathbf{p}(n)}{\mathbf{p}^T(n)\mathbf{p}(n)}e(n). \tag{4.159}$$

This implies that the direction of the update of the APA comes from the orthogonal projection of the most recent input vector $\mathbf{x}(n)$ onto the orthogonal subspace spanned by the columns of $\tilde{\mathbf{X}}(n)$ (the previous $K - 1$ input vectors).

In order to show how this fact is related to the decorrelating property of the APA, let the input signal come from an autoregressive process of order $K - 1$ [39], i.e.,

$$\mathbf{x}(n) = \tilde{\mathbf{X}}(n)\mathbf{a} + \tilde{\mathbf{v}}(n), \tag{4.160}$$

where $\mathbf{a} = \begin{bmatrix} a_1 & a_2 & \dots & a_{K-1} \end{bmatrix}^T$ are the autoregressive process coefficients and where the L dimensional vector $\tilde{\mathbf{v}}(n) = [\tilde{v}(n) \ \tilde{v}(n-1) \ \dots \ \tilde{v}(n-L+1)]^T$ contains the samples of the white noise exciting the autoregressive process. The least squares (LS) estimate (see next chapter) of $\mathbf{a}$ at time n is[21]:

$$\hat{\mathbf{a}}(n) = \left[\tilde{\mathbf{X}}(n) \right]^{\dagger} \mathbf{x}(n) = \left[\tilde{\mathbf{X}}^T(n) \tilde{\mathbf{X}}(n) \right]^{-1} \tilde{\mathbf{X}}^T(n) \mathbf{x}(n). \tag{4.161}$$

Then, the direction of the update of the APA can be written as $\mathbf{p}(n) = \mathbf{x}(n) - \tilde{\mathbf{X}}(n)\hat{\mathbf{a}}(n)$. From (4.160), this means that $\mathbf{p}(n)$ is an estimate of the white noise $\tilde{\mathbf{v}}(n)$. Hence, instead of using the correlated signal $\mathbf{x}(n)$ to do the update, the APA uses the uncorrelated one $\mathbf{p}(n)$. Therefore, an improvement on the speed of convergence can be expected. This reflects the most important motivation for using the APA: to obtain a fast convergent algorithm when the signals involved in our problem are highly correlated.

If μ is different from one, it can be associated to the error signal, so the conclusions made above can still hold (the error signal is passed through a filter with a single coefficient equal to μ). To read more on filtered error algorithms, see [55].

Finally, we can also find the APA as an approximation to an NR algorithm in the same way we did for the NLMS. However, in this case, instead of using the instantaneous estimates given by (4.2) to replace the statistics used in the NR algorithm, we use the following:

$$\hat{\mathbf{R}}_{\mathbf{x}} = \frac{1}{K} \sum_{i=n-K+1}^{n} \mathbf{x}(i)\mathbf{x}^T(i) \text{ and } \hat{\mathbf{r}}_{\mathbf{x}d} = \frac{1}{K} \sum_{i=n-K+1}^{n} d(i)\mathbf{x}(i). \tag{4.162}$$

The use of the last K input regressors into these estimates will improve the quality of the estimators. The higher the value of K, the closer the estimators would get to the actual statistics. This would make the algorithm closer to the NR operation mode, so we should expect an improvement in the speed of convergence.

4.7 Simulations Results

In this section we provide some simulations to show some of the results of the convergence analysis introduced in this chapter. Regarding the stability results, the reader is referred to [37], where extensive simulation analysis was conducted on the stability of the LMS and SDA.

In our simulations we consider a system identification scenario with the linear regression model (2.11). The input sequence $x(n)$ was generated with a Gaussian AR1 process with a pole in a, i.e.,

---

[21] Notice the similarity of this with a linear prediction problem from Sect. 2.5!!

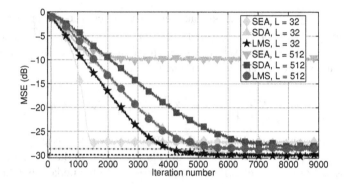

Fig. 4.7 Comparison between LMS, SDA and SEA with filter lengths $L = 32$ and $L = 512$. $\mu = 0.001$ and $a = 0$

$$x(n) = ax(n-1) + u(n), \qquad (4.163)$$

where $u(n)$ is a zero-mean white Gaussian noise process with variance σ_u^2. When, $a = 0$, the process $x(n)$ will also be white noise. As a increases towards 1, the signal becomes more colored. The system $\mathbf{w}_T$ is taken from a measured impulse response [56, Fig. 3]. In the scenario where $L = 512$, the response was truncated, whereas when $L = 32$, the samples 108–139 were used. The adaptive filter length is set equal to L on each case. The output of the system was then contaminated with $v(n)$, a zero-mean white Gaussian noise process with variance σ_v^2. This noise was computed to achieve SNR = 30 dB. The curves are the result of the averaging of the instantaneous values $|e(n)|^2$ over 1,000 realizations of the model (2.11) (using different realizations of the noises $u(n)$ and $v(n)$ in each of them). The initial adaptation of the algorithms is done in "real-time", meaning that the samples are used as they arrive. Therefore, in the first iteration the input vector is $\mathbf{x}(0) = [x(0), \, \mathbf{0}_{1 \times L-1}]^T$ and so on. It is only after L iterations that the input vector will be "full", so this can have an impact in the shape of the curves during the initial phase of the adaptation.

We start by comparing LMS, SDA and SEA. In Fig. 4.7 we set $\mu = 0.001$ and $a = 0$ for all the cases. We used $L = 32$ and $L = 512$ to study the effect of the filter length on the convergence behavior. When the filter is short, the SEA converged rather fast but with a larger EMSE than the other algorithms. However, the SEA performance is severely compromised with the larger filter. In order to achieve a smaller EMSE (closer to the one of the other algorithms) the value of μ should be decreased. If we do that, the problem is that the speed of convergence will be worse as we have already seen in Fig. 4.4. The LMS and SDA have almost the same steady states (dashed lines) as the predicted using (4.133) and (4.147). In both cases, the EMSE increases with L. In terms of speed of convergence, the LMS is faster than the SDA (the price to pay for quantizing the input data), and both get slower with larger filters.